The Rocket Company

by

Patrick J. G. Stiennon

and

David M. Hoerr

Illustrated by

Doug Birkholz

American Institute of Aeronautics and Astronautics, Inc.
1801 Alexander Bell Drive
Reston, Virginia 20191–4344
Publishers since 1930

American Institute of Aeronautics and Astronautics, Inc., Reston, Virginia

1 2 3 4 5

Library of Congress Cataloging-in-Publication Data

Stiennon, Patrick J. G.
The rocket company / by Patrick J. G. Stiennon and David M. Hoerr;
illustrated by Doug Birkholz.
p. cm.
Includes index.
ISBN 1-56347-696-7
1. Launch vehicles (Astronautics)--Design and construction. 2. Reusable space vehicles. 3. New business enterprises--United States--Finance. I. Hoerr, David M. II. Title.

TL785.8.L3S84 2005
629.44'1--dc22

2005016557

Cover design by Aaron Bertoglio.

The book is dedicated to those who have devoted their lives and fortunes to opening up the final frontier—the rocket pioneers of the first years of the 20th century, the founders of early companies like Reaction Motors, the launch vehicle entrepreneurs of the past 25 years, the X PRIZE contenders, and all those currently working to build suborbital and orbital transportation systems.

Foreword

There is a large and vibrant marketplace of individuals willing to pay for the opportunity to fly into space. Recent surveys consistently indicate that more than 60% of the U.S. public would welcome the opportunity to take such a trip, and nearly 13,000 consumers per year would spend upward of $100,000 for a flight. For the first time, a billion dollar market exists outside of the government which involves the flight of humans into space. Similar to the birth of aviation and the start of the personal computer market, the personal spaceflight revolution holds out the promise of low cost and high reliability space travel for the masses. The only thing missing has been the privately owned and operated spaceships needed to serve this market.

On 18 May 1996, under the St. Louis Arch, I announced an idea that would spur the industry into motion to develop these private spaceships. Joined by the St. Louis business leaders, NASA Administrator Dan Golden, and 20 veteran astronauts, we offered $10 million to the first private team to build and fly a spaceship carrying three adults to 100 kilometers altitude, twice within a two-week period. Our primary goal was to attract and motivate entrepreneurs, risk-takers, and innovators to develop low cost and innovative approaches to space travel. More specifically we wanted a new generation of spaceship designers, not the traditional aerospace primes who were locked into expensive and unimaginative programs. The concept worked. Within a few years, the Ansari X PRIZE had registered 26 teams from 7 nations willing to compete.

On 4 October 2004, Burt Rutan and his Mojave Aerospace team, supported by private financing from Paul Allen, won the Ansari X PRIZE Competition. The result of the X PRIZE flights was a miraculous rise in the public's demand for spaceflight, coupled with the private sector stepping forward with private funding to develop the vehicles. For the promise of a $10 million prize, more than $50 million was spent by the competing teams in research, development, and testing. Dozens of real spacecraft were actually built and tested. Compare this to a $10 million investment from a government procurement program, which historically has resulted in one or two paper designs.

This is Darwinian evolution applied to spaceships. Rather than paper competition with selection boards, the winner was determined by the ignition of engines and flight of humans into space. Best of all, the X PRIZE Foundation didn't pay a single dollar until the results were achieved. The bottom line is that prizes work!

Entrepreneurs can solve the problems that large bureaucracies cannot. Prizes can result in fixed-cost science and fixed-cost engineering. More importantly, prizes can harness the passion and dedication of the entrepreneurial mind that cannot be purchased at any price.

As a result of the ANSARI X PRIZE Competition, the front pages of *Forbes, Investors Business Daily, The Wall Street Journal, Wired, The Washington Post,* and *The New York Times* began to report on a new breed of space entrepreneurs. Companies representing the X PRIZE teams, XCOR, SpaceX, Zero Gravity Corporation, and Space Adventures captured both public attention and investor interest. For our space community, these companies were the early versions of Apple©, Microsoft®, and Netscape®. These companies embodied the entrepreneurial "can-do" spirit of America.

Most of the new space companies are focused on one specific market: personal spaceflight. Many of us believe that it is the only commercial market that makes near-term sense. Call it space travel or barnstorming, the fact is that the public will pay for a chance to fly into space. This is a mass market that can yield a profit while developing breakthroughs in launch operations. These two areas are the very essence of what is most needed to develop a hearty industry.

The reason that spaceflight has historically been so expensive is simple—there just isn't enough of it. The commercial launch market for satellites is pathetically small, only 15–25 launches per year, worldwide. The number of human space launches is even smaller, only four space shuttle flights and four Soyuz flights in a good year.

What we need is not dozens, but thousands of spaceflights per year. Flights that teach us about launch operations—how to refuel, retool, and relaunch a fleet of reusable vehicles.

I recognize that the vehicles resulting from the X PRIZE are only suborbital ships, only one-fortieth the energy requirement of an orbital ship, but the lessons we will learn from operating these vehicles are critical. We will learn about operations, an area in which we are sorely lacking.

Everyone knows that the reason the space shuttle costs so much to operate is not the fuel, but its dependence on a standing army of 10,000-plus professionals. We have people, watching people, watching people in order to increase safety margins.

In stark contrast, the reason that a crew of 6 can turn around a Boeing 737 for its next flight in 20 minutes is the operational robustness achieved through millions of flights conducted during the first 50 years of aviation. Flights that began with 10-minute hops across farmers' fields grew over time to transatlantic journeys. Our space program has in essence skipped the learning stages of these 10-minute hops and went straight to orbital shots. We need to practice and learn, but we cannot achieve the flight rates and experience base we need with the space shuttle or the Crew Exploration Vehicle or any other large government program.

The next generation of X PRIZE vehicles will soon be competing in the X PRIZE Cup—an annual competition for rocket-powered aircraft and future spacecraft. The X PRIZE Cup is a partnership established between the X PRIZE Foundation and the state of New Mexico under the vision of Governor Bill Richardson—specifically to support the new generation of space entrepreneurs.

Finally I would like to address the issue of risk. In contrast to individuals who speak about reducing risk, I want to speak in favor of taking more risk. There is no question that the ANSARI X PRIZE Competition involved risk—so does going to the moon or Mars or opening any portion of the space frontier. But, this is a risk worth taking! As Americans, many of us forget the debt we owe to early explorers. Tens of thousands of people risked their lives to open the "new world" and the American West. Thousands lost their lives crossing the ocean and then the plains—but we are here today because of their courage.

Space is a frontier and crossing new frontiers is inherently risky! As explorers and as Americans, we must have the right to take risks that we believe are worthwhile and significant. We owe it to ourselves and to future generations. It is also critical that we take risks to develop technology. It is critical that we allow for failure. Without risk and without failure, we cannot initiate and realize the very breakthroughs we so desperately need to open the space frontier.

A breakthrough, by definition, is something that was considered a "crazy idea" the day before it became a breakthrough. If it wasn't considered a crazy idea, then it really wasn't a breakthrough, but an incremental improvement. Remember Gene Kranz' immortal words, "Failure is not an option" . . . if we live and work in an environment where we cannot fail, then breakthroughs may not be an option either.

In summary, I urge the reader to support those efforts that will allow us to realize our dreams of space exploration. Support prizes as the most efficient means to foster and enable breakthroughs in technology and embrace risk. Help the American people understand that space exploration is risky—but a risk worth taking. Let's let space explorers be heroes once again.

The Rocket Company comes as a needed antidote, at a time when economic interests are driving companies to consider their future plans as valuable proprietary assets. *The Rocket Company* lays out a highly detailed engineering plan and business model bridging the gap between suborbital passenger flight, and the colonization of Mars. The book is a highly accessible—almost a textbook or case study, but more engaging—realistic and believable scenario for achieving low-cost space transportation to orbit and beyond. The combination of an accessible, entertaining, yet thorough examination of the questions which must be addressed, by the post-X PRIZE entrepreneurs makes the book a must read for all engaged in developing lower-cost space access and developing new markets in space. *The Rocket Company* should serve as a point of departure for the ongoing development of the growing consensus, which the X PRIZE has created, and which has been so effective in motivating progress.

Peter H. Diamandis, MD
Chairman/CEO, X PRIZE Foundation
Chairman/CEO, Zero Gravity Corporation
Founder, International Space University
Santa Monica, California

TABLE OF CONTENTS

Acknowledgments

The purpose of *The Rocket Company* is to contribute to and encourage the ongoing debate about how low-cost access to space can be achieved. Although the book is a work of fiction, its intent is to make available lessons learned by the authors on the job and by following for many years the history and travails of various efforts aimed at achieving lower cost space transportation. Any errors and omissions are purely the responsibility of the authors. Special thanks to David P. J. Stiennon for transcribing the first draft, to Clark Lindsey for publishing an earlier draft of the book on HobbySpace.com, and to the following people and organizations for advice or comments on some part of the text: Pratt & Whitney Space Propulsion, Dennis Haas, Constance Nielsen, Robert Stockman, Bob Schofield, John Routledge, Christopher D. Hall, Dick Morris, Cathy Palmer, Bob Witt, and Wini Hoerr. Finally, we would like to acknowledge the debt owed to those for whom or with whom we have worked, especially Gary C. Hudson, Maxwell W. Hunter II, Lee Lunsford, Eric F. Laursen, Tom Brosz, James R. Grote, and Paul Bridenbaugh.

CHAPTER 1
Seven Billionaires and One Big Problem

JOHN Forsyth is a man with a net worth of approximately one billion dollars. Like many others in the last decades of the last century, he had made his money in the software industry. In the late 1990s he had also, for primarily ideological reasons, funded a small company comprised of space enthusiasts who believed that they knew how to reduce the cost of access to space by a factor of 10 or more. It was Forsyth's belief, and one shared by many, that for all of the exciting visions of humanity's future in space the essential enabler was lower transportation costs. At that time his personal assets were considerably smaller, and after expending a large portion of his net worth to develop a new space launch vehicle, they needed still more money to take the project to completion. It soon became apparent that the usual sources of additional funding, especially Wall Street, were not yet convinced that rockets were a viable investment. At the same time, there were problems with the design. The end of the venture produced no clear technical consensus of how to achieve the goal of low-cost space launch. A lot of interesting technological experimentation and simulation was done, but in the end no clear blueprint for achieving success had been developed.

Thus Forsyth's first foray into the business of building rockets left him convinced that, before investing in space launch again, he had to have two things firmly in hand: a clear plan for making money, in order to attract the large amount of capital that his experience showed would be needed, and a clear technical plan or approach for building the rocket, so that success could reasonably be assured.

On the technical side, NASA and the space agencies of both major and minor world powers had shown that big budgets of hundreds of millions to tens of billions of dollars would not by themselves produce any clearer consensus as to how transport to Earth orbit could be most economically achieved. He had carefully considered not only his own failed effort, but also earlier private efforts starting with the German company OTRAG back in the mid-1970s. It was clear in his mind that it would take more than merely private initiative, however much that was needed, to produce any major leap over the inability of governments to bring down the cost of space transportation.

I first met Forsyth at a conference of the L5 Society in San Francisco in 1984 and had bumped into him from time to time since then at various gatherings of the space "faithful," those who believed that man (the term still meant humanity in those days) had a destiny that included moving into space and occupying at least the solar system. Forsyth was one of those entrepreneurs who had in mind making a great deal of money at least in part so that he could achieve something in space. Forsyth, although certainly not the richest man to come out of the last decades of the last millennium, was one who had amassed a considerable fortune. At the same time, he was not particularly impressed by that wealth.

He now was proposing to risk it all if a real chance for opening up the space frontier could be found. My own background as a science reporter and author of several books detailing the successes and failures of various technical enterprises led him to engage me to observe and record the story of what he hoped would be a great adventure. My books tended to focus on the adventure inherent in bringing to market new technologies or describing the excitement present even in mundane industries. Building rockets, however, has never suffered from being perceived as mundane. If nothing else, their unpleasant habit of blowing up—which they still did with some regularity more than five decades after the dawn of the Space Age—earned the industry a certain notoriety. Even the popular phrase "it's not rocket science" paid tribute to a universal understanding that rockets were still a domain of technology needing all of the brain power that could be brought to bear.

Building rockets, however, has never suffered from being perceived as mundane.

But the real problem in reducing the cost of space launch, as Forsyth had explained to me, was not the technology; much technology development had been accomplished with those billions that had been spent in the government programs. No, the real problem was in finding a way to make money in the transportation side of the space industry, so that investors would be inclined to make the necessary capital available. The satellite business—especially the geostationary communication satellite segment—had proven to be extremely profitable, although with the useful orbital slots nearly full it was now in a replacement-and-improvement mode with little growth in new systems or players. Even the low-Earth-orbit satellite constellations looked like they were beginning to make money.

The same couldn't be said for space transportation. The most reliable of the expendable launchers—supplied by Europe, the Japanese, the Russians, the Chinese, and in the United States, Boeing and Lockheed Martin—were used with some regularity to launch the comsats and occasional scientific spacecraft. But for the most part, they managed to avoid losing money only because of government subsidies, either directly or through the support infrastructure at the national launch ranges. The Russians, even with their impressive rocket engine technology, still lacked adequate resources to capitalize on their abilities. Various U.S. companies had purchased Russian technology and were still cooperating with them—mainly by paying royalties, but occasionally by engaging in joint production activities. Meanwhile, it seemed that every moderately developed nation in the world had developed one or more expendable launch vehicles and was aggressively marketing them.

FORSYTH was politically a libertarian, in the old sense of the word. He was also an adherent of the Turner theory of the American frontier. Frederick Jackson Turner, a turn-of-the-century historian from Wisconsin, had set forth the thesis that what made America free was the open frontier. America had become a lot less free in Forsyth's reckoning. But if you believed in Turner's thesis, a new frontier would make even those remaining at home a lot freer. Forsyth was among those who saw the Internet as a kind of virtual frontier that had been largely responsible for the resurgence of democracy, which had followed the successful conclusion of the Cold War. It also served as an economic frontier, the likes of which had not been seen since the railroad boom of the late 19th century. For awhile the growth of the Internet economy had seemed to be outrunning even the endless growth in the federal government and its budget, but the long boom of the 1990s had finally ended with a shock whose effects were still being felt. It was, after all, only a virtual frontier, and the pioneers and "settlers" were still bound by the constraints of the "Old World."

Forsyth thought that America—with all of its problems—was still the best place in the world to live, and his loyalties were strong. But the trends were troubling. Budget deficits were growing again, and a new value-added tax with its inevitable higher tax burden was beginning to seem likely. At the same time, the stifling effects of endless regulation and new social programs made it seem certain that America was heading where Europe had already gone: high unemployment, heavy-handed bureaucracy, and rapidly aging population. The latter was a problem because an aging population meant, obviously, fewer young people to provide the stimulus as well as the hands and brains for any kind of progress. At home, the population was aging too, but it had not yet reached crisis proportions, thanks to immigration both legal and illegal. But rising unemployment and the cost of social programs fueled a growing political consensus against immigration in the United States.

TO be revitalized, America needed a new frontier, in John Forsyth's opinion. A new frontier meant a place where ambitious men and women and capital could go if—or when—the environment at home became too oppressive.

Among the entrepreneurs who made their fortunes during the last decades of the 20th century, Forsyth was not alone in being seriously interested in space exploration. Forsyth had gathered together a group of successful entrepreneurs, each with a substantial fortune, who shared his interest in space. The reasons for their interest in space were almost as varied as their number, and because many of them had more than one motive, all of the motivations exceeded the number of prospective participants. From this pool of wealthy space investors, Forsyth had formed an investment group. Their motives ranged from the ridiculous to the sublime, from a desire to contact extraterrestrial intelligence, to building a Benedictine monastery at the lunar South Pole. A few members of the group were still interested in proving that they could take on the next technological frontier and make money doing so. But every member of this small group had been profoundly affected by the early days of the Space Age, from Sputnik to the end of Skylab in 1975.

They had thrilled at the Apollo lunar landings and had marveled at the streams of pictures brought back by the space probes of the 1970s, 1980s, and 1990s. But the momentum of the early days dissipated as Apollo was shut down and the nation's human spaceflight efforts were narrowed to a single vehicle, the space shuttle. But the shuttle never did provide the spark necessary to set off a new age of space exploration and development. The excitement was gone, and the spirit of adventure died. But not the longing. These six men and one woman had made themselves a promise in those halcyon days, like so many others, that they were going to personally take at least one trip into orbit, if not to the moon. Unlike so many of their contemporaries, they really meant to fulfill that promise, and not by bribing some government to stuff them into a capsule for a few unpleasant days in space. No, they meant to go in style, like the space travelers in the movie *2001—A Space Odyssey*.

Of course, they hadn't reached their positions or their fortunes without a keen sense of practical realities. They were not interested in adding yet more billions to the private and public monies that had been lost in pursuing the goal of low-cost access to space. They were all familiar with (and in a few cases had been themselves among) the multimillionaires and billionaires, who had invested in space transportation start-ups in the past 30 years. Even with the help of some of the "fading lights"—experts from the glory days of Apollo—these efforts had had no practical effect. Space transportation was still as costly and far from routine. The human race was still planet bound, save for the handful of government astronauts or the even tinier number of rich tourists who were dependent on the space shuttle and the venerable Russian Soyuz for their short and infrequent flights to low Earth orbit or the International Space Station. The promising suborbital tourist market, which was demonstrating some success, was still a long way from routine access to orbit and the solar system beyond.

WHY did spaceflight still cost so much? As far back as 1952, Wernher Von Braun had proposed a very ambitious Mars Project. He estimated that a three-year Mars expedition with 10 vessels, 70 men, and placing 600 tons in Mars orbit would have logistics requirements no greater than those of a minor military operation in a limited theater of war. Von Braun's approach to cost estimation was to look to the propellant requirements, which he could calculate fairly accurately, as a good measure of the total cost. He estimated the cost of the propellants at approximately $500 million, or about $100 per ton and assumed that would approximate the total cost. A key supporting assumption was that the launch vehicles used to mount the expedition would already be developed, with a demonstrated reliability comparable to that of modern transport aircraft. That last part, as history had shown, was the real trick!

The cost of the highest-energy propellants—liquid oxygen (LOX) and liquid hydrogen (LH_2)—these days is about $400 per ton, and less energetic propellants such as liquid methane and liquid oxygen cost less than $100 per ton. When inflation is taken into account, the real costs of propellants today are less than Von Braun assumed by a factor of three to ten. The Saturn V launch vehicle, which Von Braun developed for the Apollo program, required about 22 pounds of propellant for each pound of payload launched into orbit. The Saturn used LOX and LH_2 in its upper stages, and lower-energy propellants consisting of a rocket-grade kerosene and liquid oxygen in its first stage. Even at $400 a ton, the cost of the propellants required to place a pound into low Earth orbit is less than $10,000 a ton, or $5 a pound. However, as Forsyth was all too aware, the market price for placing a pound into Earth orbit was over $3000, and much more if some of those pounds were people.

The recent history of the space program in the United States was full of promises that the next new launch vehicle program was going to reduce the cost of space transportation by a factor of 10, 8, 4, or 2, but none ever had. It seemed impossible to validate Von Braun's assumption that you could build a space launch vehicle that had the reliability and operating cost parameters of a modern transport aircraft. It had been over 50 years since Von Braun had proposed that such a thing was possible, and hundreds of billions of dollars had been spent on space transportation without significantly reducing costs. One could argue that the Soviets and the Chinese had achieved some economies of scale in the quasi-mass production of their expendable launch vehicles, but it was difficult to calculate true costs within command economies. Still, neither the Chinese nor the Russians nor anyone else were selling launch capability significantly below market prices, and given that the market was considerably overbuilt, it seemed likely that $3000 per pound was close to cost, if not less than cost for most providers.

FORSYTH had long been part of a Los Angeles-based Camerata, which had gathered quarterly for almost two decades for the purpose of exploring ways and means of opening up the space frontier. The members were unanimous in their belief that this was utterly dependent on a major reduction in the cost of transportation to orbit. Therefore, they studied, critiqued, and learned from the various new launch vehicle companies that were forever being created only to fail a few years later. Many of these launch vehicle companies had been formed or funded by Camerata members who, after expending anywhere from $20 million to more than $100 million, had had to pull the plug on the ventures. A few of these companies had been technically successful, placing payloads in orbit, or carrying a few passengers on suborbital flights, and had gone on to become small players in the aerospace industry. None of them, however, had lowered the cost of space transportation enough to really change things. Given the incredible cash "burn rates" necessary for any serious dabbling in rocketry, in each of these cases it had soon become clear that with the relatively small amounts of money available, a revolutionary reduction in space launch costs was out of reach.

The kindred souls which Forsyth had assembled, most of whom he knew from the Los Angeles Camerata, comprised seven individuals with a combined net worth of approximately $10 billion. Of that sum, $2 to $5 billion were potentially on the table for this latest venture. They were by no means intending to squander yet more billions in a quixotic space enterprise. Rather, they were united by both a strong desire to bring about low-cost space transportation and open up a new frontier, and a by firm commitment to make another attempt only on the basis of a well-conceived plan that included a real probability of success—both technically and financially. That was absolutely essential before any of them would commit any significant amount of money.

> The opening of the space frontier just needed the appropriate economic engine to drive it.

He did not have a plan yet, but Forsyth was a believer. He was convinced that the physics was there, that it could be done technically. The opening of the space frontier just needed the appropriate economic engine to drive it. Bill Gates always talked about creating a positive economic spiral for a new product or service; a favorite example was the CD-ROM as a medium of information storage. If there were large numbers of useful products on CD-ROM, then consumers would demand CD-ROM drives, and they would be installed on every PC. On the other hand, if CD-ROM drives were installed on every PC, developers would make the investments to come up with new applications. To get the process started, Microsoft had invested heavily in creating CD-ROM products, with their Encyclopedia Encarta being the most significant. Once the process was started, it fed on itself: more installed CD-ROM drives meant that the cost of the hardware dropped, and once applications developers had a bigger market, the cost of those applications fell as well.

Forsyth felt sure that it could work for space transportation, too. Lower transportation costs would result in a larger demand for transport to orbit, and a greater demand for transportation would result in lower operating costs and improved operating efficiencies, which would further lower space transportation costs, which would further increase demand for transportation in an ever-growing spiral. No one, certainly not any of the government space agencies, had ever come close to getting a self-reinforcing spiral of lower cost and greater demand started in space transportation.

In the early days of the space shuttle, NASA had attempted to provide low-cost transportation to orbit in the form of getaway specials and other small payload flight options, but these had been made available more for political reasons than for any real belief that demand could be stimulated. Furthermore, the cost of interfacing a small, inexpensive payload with a multibillion dollar national asset, and the many years of delay before your payload was finally launched, had always kept the real cost high.

On the other side of the cost equation is demand. The combined NASA and U.S. Air Force budgets for space transportation (excluding ground infrastructure) totaled $6 to $7 billion a year; worldwide commercial demand only another $4 to

$5 billion, and other national space programs an additional $5 billion. This budget of $15–17 billion per year for space launch services resulted in less than the equivalent of 1.5 million pounds a year transported to low Earth orbit. If costs dropped to $500 per pound, a seemingly reasonable near-term target, total weight to orbit just for the commercial demand would have to increase to 4 million pounds, just to keep the market at the size of the $2 billion that Forsyth thought was the minimum to support the necessary investment. It would be a long wait to develop a demand of 4, 8, or even 10 million pounds to orbit, which would be necessary if a reasonable return on a not-insignificant investment was to be achieved. Hence, the few new companies that had actually managed to fly payloads and satellites into orbit had to charge near-market rates to keep from running out of cash and were unable to stimulate that much new demand.

It seemed to Forsyth that in order to get the growth spiral started it would have to become obvious to everyone that the cost of space transportation was on a really steep downward spiral. Something like Moore's Law would have to be set to work, with costs falling by half every 18 months or so, for the foreseeable future. Forsyth had researched and studied the problem long enough, including talking in depth to as many knowledgeable experts as he could find, to be firmly convinced that no laws of nature would have to be overturned and no new science would have to be invented to achieve at least the first factor of 100 reduction in space transportation costs. All that was missing was the right engineering. And the right business plan. The big question was how much it would cost, and whether he could bring the necessary resources to bear.

The Group of Seven, as the members of the investment group had taken to calling themselves, had now been meeting regularly for over a year, and Forsyth had actually been able to get them all on the same page. If workable business and engineering plans could be developed, he felt confident that they would commit to a major investment. A limited liability corporation had been formed, called AM&M, and also a limited liability partnership so that they could funnel money into the development program while keeping open the possibility of benefiting from tax-deductible losses. To bring the necessary technical team together, an initial round of $50 million had been invested through the limited liability corporation. The task now was to put together a business plan, the most important part of which would be a marketing plan that showed how the company would actually make money. That was the tough nut to crack. All of the technology and engineering in the world wouldn't be of any use if they could not demonstrate that the enterprise would pay off.

Although the marketing plan was the first major project, the first step was to hire a small engineering team so that marketing ideas and strategies could be checked for technical sanity. Secondarily, the engineering was going to have to be absolutely first rate, and that meant pulling together a real team, not just a group of employees. The longer the team was together before they got all of the way into a full-blown vehicle development program and real money was being spent, the faster and more smoothly the project should go.

CHAPTER 2
Big Telescopes, Hot Rodders, and Librarians

JOHN Forsyth had been keeping track of people who had actively worked on the more interesting aerospace vehicle programs in the last 10 to 15 years. His choice to head up this project was an engineer named Tom Rabbet, currently employed at Boeing Aerospace. Tom had recently worked on the Advanced Launch System (ALS)/Delta derivatives, and also on the SeaLaunch system. Boeing's joint project with Russia to create the sea-based launch system had given Boeing—and Tom—some real experience in vehicle design and vehicle operations, which was perhaps more important. Tom was of the right age, about 45 years old, seasoned but energetic, and experienced in successfully managing an engineering group. He had the good fortune to have spent most of his career designing and sometimes even building pieces of launch vehicles. He knew what had to be done in order to check the feasibility of a particular technical solution, a design approach, or even a complete vehicle design.

Tom quickly got to work on hiring his core team. Based on his extensive networking throughout the aerospace industry, he was familiar with those engineers who really knew their stuff and had a deep intuitive grasp of their disciplines. For structures, he chose Lester Weld from Boeing; a thermodynamicist from Lockheed Martin, Sol Meir; from Pratt & Whitney, Chet Lovell, a rocket propulsion engineer; guidance and control man George Poindexter from McDonnell Douglas; a performance analyst, Paul Reston, who had left the aerospace industry more than a decade earlier; a cryogenics engineer, Jessica Durbin from Liquid Air; and finally, a weights engineer, Dave Foreman, who had worked for General Dynamics before it was acquired by Martin and subsequently by Lockheed and whose experience with the weights of all sorts of launch vehicle and spacecraft components went back to the Apollo days. Aside from Dave, who was over the retirement age, and George, who was 35, everyone was about Tom's age, in their mid-40s.

The budget for the engineering team was $25 million a year, which could support about 100 engineers, if they didn't do too much actual "tin bending," or hardware building. The initial recruiting drive drew heavily from the ranks of younger engineers—some of them just out of college. They were certainly chosen for their smarts, but not the purely analytical type. What Tom needed were people who had a record of designing and actually building things—people who knew how to come up with clever and clean designs and then make them work. The hiring was slow and deliberate. Tom had read Bill Gates' biography and tended to agree that, for a business, the most expensive mistake was hiring the wrong people.

Tom also had a theory that one problem with modern engineering schools was that they tended to turn out engineers who thought of themselves as failed scientists. He believed that engineering was very different from science. Engineering was way older. The Romans had built engineering marvels that were still standing after 2000 years, using technology that had been developed more than 1500 years before the dawn of modern experimental science. Engineering understanding was good enough, when it allowed you to build something, and typically the engineers got there first. For example, engineers had been using complex numbers to solve real electrical engineering problems long before the mathematicians were ready to believe in the square root of −1.

The Romans had built engineering marvels that were still standing after 2000 years, using technology that had been developed more than 1500 years before the dawn of modern experimental science.

Tom's favorite engineering story concerned the construction of the Hale 200-inch telescope at Mt. Palomar, best told by Richard Preston in his book *First Light*. Building a telescope that big that would actually work was not technically possible in the 1940s when it was proposed. But Hale—who claimed to have had numerous meetings with an imaginary green elf—had the motivation and the charisma to get it built anyway. The problem with such a large telescope mirror was that the weight of the mirror itself would cause it to deform when it was moved for pointing, so much so that it would be impossible to bring the telescope into focus. However, a certain clever Caltech engineer designed and built 50 mechanical computers that were driven by pendulums hanging beneath the telescope mirror. As the telescope was rotated to point to various positions in the sky, the pendulum weights would swing as a result of gravity and drive the pendulum computers to press against the back of the mirror with precisely the right amount of force to keep the mirror shape perfectly parabolic. The computers were way beyond the technology of the times. They never needed repair and functioned nearly flawlessly for 50 years. Occasionally they would get stuck, but swinging the telescope back and forth from side to side a few times always fixed the problem.

The Russians didn't happen to have such a genius on hand when they built their 6-meter telescope in the 1970s, and so never got it to actually work unless it was pointing straight overhead. It wasn't until the early 1990s that new technology in the form of high-speed digital computers became available to replace the pendulums and mechanical computers.

That kind of genius is hard to duplicate. But science is repeatable; anybody can repeat an experiment. Engineering, though, often goes beyond what science can do, and sometimes a few people in one time and place can accomplish what no one else can. The SR-71 was another example. It probably wasn't technically feasible at the start of the program. But somehow Kelly Johnson and the Skunk Works had made it happen anyway. No aircraft matched its capabilities for more than 40 years after it had first flown.

Tom didn't think that the design of the first launch vehicle capable of routine, low-cost operations was going to require the SR-71 sort of doing a little more than was really possible. Still, it was clear to anyone that if such a vehicle could be built with current technology it was going to take good engineering—very good engineering. Tom's experience at Boeing had certainly taught him that it was a rather small percentage of the engineers who had most of the new, good ideas. Therefore he looked for bright engineering types who really did things. Tom wanted something beyond brains; he was looking for people who built things that worked—boats, cars, hot rods, shareware computer programs, special-purpose computers, home-built airplanes, people who took their senior design courses so seriously that they really built something that worked, and even the occasional backyard rocket builder. A long list of patents on the resume didn't hurt, either.

IN addition to the engineers, Tom's first hires included a couple of librarians. Well over a hundred billion dollars had been spent on rocket and launch vehicle technology in the United States during the past 50 years, mostly by government agencies. It was therefore fairly well documented in various archives. AIAA and the International Institute of Electrical and Electronics Engineers (IEEE) were two professional societies that had their own well-organized volumes of relevant technical papers, many of which were available on CD-ROM or online. Information from the European, Japanese, Chinese, and Russian space programs was also available, but was progressively more difficult to access. The Russians were still hungry for hard currency, though, and so it was possible to buy access to most of their technology.

Well over a hundred billion dollars had been spent on rocket and launch vehicle technology in the United States during the past 50 years, mostly by government agencies.

It was Tom's intent to gather the best of this information and then arrange and index it in a way that would facilitate a thorough assessment of past design solutions to the various problems that would be encountered in designing the

American Mining and Manufacturing (AM&M) launch vehicle. The primary purpose of this technical library was not to gain access to technology; after all, if you wanted technology that had been proven to work, you went directly to whichever vendor was providing it to the marketplace. The real purpose of the library was to get access to the experience and understanding that countless thousands of engineers had gained over many years of toiling in the bowels of various government space agencies or their industrial contractors, sometimes at the cost of life and limb. During slow times, when there was no money to spend on new projects, government engineers had written design manuals and handbooks that summarized the history and engineering experience gained in earlier projects. Much of this information was still available from the National Technical Information Service.

The library, as it came together, was physically laid out like a two-stage launch vehicle. (Tom was reasonably certain that the vehicle they would end up building would have two stages.) Therefore, all material pertaining to first-stage engines, for example, was located together, rather than being kept in an all-encompassing propulsion section that also contained information about upper-stage engines and attitude control thrusters. The idea was to organize the material so that the engineers would have all of the information relating to whatever subsystem of the vehicle they were working on right next to the material that dealt with the other subsystems that their part would be expected to accommodate. The librarians, with years of experience in technical information management, ensured that the layout made sense, and also built an easy-to-use computer-based catalog and database containing most of the library's content.

Jeff Bezos at Amazon was working toward having every book, magazine, dissertation, or paper that had ever been published word searchable at and buyable on the web. However a lot of the aerospace material was not yet available, and given the relative obscurity of many of the references it might be many years before the universal library was complete. However part of the database did consist of vetted data searches that linked to the ever-growing universal library which was Amazon.

An interesting note about aerospace technical documentation was that—after about 1963—nobody wrote design handbooks anymore; rather, they wrote computer programs. And although some of the old programs were still running at the big, old aerospace companies like Lockheed Martin and Boeing, it was very hard to gain a real understanding of the technology and the design insights that were buried in a computer program. It was much easier to read and comprehend a well-written handbook than to understand the thinking behind a few thousand lines of computer code. And usually that code was poorly documented.

THERE are some aspects of engineering that, although perhaps not unique to the aerospace industry, were in Tom's opinion much more prevalent there—flights of technical "insanity" and stubborn, arrogant, pride. The way Tom saw it, aerospace was particularly prone to these afflictions because it was often so difficult to tell the difference between what was merely extremely difficult and what was simply impossible. So from time to time some individual or institution would, for one reason or another, espouse a technical position that just didn't make any sense, if you got down to looking at the physics involved. Because *maybe* it could be made to work, and if it did, then wouldn't that be great? Besides, if so-and-so thinks that's the way to go, well, he does know his stuff, doesn't he? And then a lot of money would be spent pursuing the latest technical wild goose. Government organizations and contractors would create whole new divisions, people would be hired, new programs would be funded, and everyone would jump on the bandwagon. Until finally, after years of futile effort, the idea would die a (usually) quiet death. In some ways, aside from the misspent resources, the worst effect was the crowding out of other sensible ideas that probably would have worked. Sometimes this resulted in people coming to disbelieve the sensible solution, while being convinced that the "wacky" solution was the only way to go! Of course, sensible but extremely difficult technical programs sometimes failed for nontechnical programmatic reasons, such as no more money or a new administration with new priorities taking office. Thus, it often took more than merely the right approach to win the day.

With the necessity of having people like Kelly Johnson go out and win government contracts on the basis of their personal reputation, pride was also an occupational hazard. To be convincing to skeptical customers, *you* had to be completely convinced that yours was the only path to success, whether or not you were right or off on one of those flights of technical insanity. Even in the lower levels of engineering, pride could be a problem. Big government programs, including just about all rocket development programs, typically employed thousands of engineers, and realistically only a few of them could have any real input. At most aerospace companies, most engineers, rather than having any real input into the design process, got used to spending their days answering stupid questions from management who did not know the right questions to ask and who were no longer interested in the answer to the question by the time the analysis was done. This also tended to encourage the development of a certain arrogance—"I am the expert, so of course I'm right and you, by definition, are wrong!"—because that was the only way you could wield any power.

If you were not too concerned about whether anyone paid attention to what you thought should be done, then aerospace could at least provide you with bright, interesting colleagues, all of the computer resources you wanted, good pay, and even a relatively low-stress job. The result was that there were a lot of frustrated engineers out there, some very bright, but a lot of them not necessarily grounded in practical design experience. The dwindling number of programs that got funded for full-scale hardware development exacerbated the problem. Fewer and fewer engineers actually got to build things. The result was a lot of junk engineering. Many papers and monographs, especially in the area of big-picture design dealing with how new problems should be approached, were a lot of nonsense because the writer had no practical experience. Likewise, there were technical fads that begot many conferences and papers, but had little practical utility. For example, endless papers were written by people who thought computational fluid dynamics was "cool," but who were oblivious to the question of how it could effectively improve the design process.

For the amount of money and time that had been spent analyzing the space transportation problem, there wasn't a lot of clear thinking. More precisely, there was some clear thinking, and an awful lot of confused thinking. The first task of the engineering team, then, was to pull together for the library as much material as they could lay their hands on that contained sound experientially based understanding of the problems faced in designing launch vehicles. In the process, they performed an extensive review of the history of the technical development of the aerospace industry. Tom made sure that this was accomplished by the very engineering team members who were going to need to build on that history. Buried in the history of any technical discipline is an insight into the technology itself, as well as an understanding of why things are the way they are. Interesting—but abandoned—innovative solutions litter the roadside of technology history.

AS in any new industry, in the early days of aerospace hundreds of ideas were tried, some adopted, and some discarded. Over time the best practical solution to a particular problem becomes the industry norm, and then "everybody does it that way." Sometimes, though, it's useful to go back and reassess the technical choices that were made and why. It is especially instructive if you can go back to the very beginnings, before people were blinded by the universally accepted approach. Considering the way that space transportation had developed—or, depending on your opinion, not developed—it was a good idea to reassess whether any wrong approaches had been taken, and where better approaches might be feasible.

Tom even had the engineers searching the online patent databases maintained by the U.S. Patent and Trademark Office. However, it was soon apparent that there was relatively little historic launch-vehicle technology preserved there. Robert Goddard was probably the last person to do substantial work in obtaining U.S. patents in rocketry until the early 1990s. This lack of technology in the patent database had surprised Tom, given the money spent and national interest in space technology in the past 60 years. However, the act that created NASA had given ownership of all patents developed with even the smallest amount of NASA money to that agency. And at the time there was no provision for private companies to obtain rights to the technology that they developed. Because almost all space technology development had been done for NASA and other government agencies, that covered pretty much everything. In fact, these patents actually issued in the name of the administrator of NASA. In some ways aerospace technology was like business methods and computer programs when it came to patents; there was very little "art" at the Patent Office. More than a few invalid patents had been issued on launch-vehicle concepts during the last 15 or 20 years simply because the Patent Office didn't have access to the history of the technology. So, although the library was meant to be primarily a design resource, it did have the additional function of providing a basis for obtaining truly valid patents and at the same time providing ammunition for shooting down existing patents that should never have been issued in the first place.

More than a few invalid patents had been issued on launch-vehicle concepts during the last 15 or 20 years simply because the Patent Office didn't have access to the history of the technology.

Tom also started a small, in-house "university." Word had gotten out that some serious money might be spent on a launch-vehicle development program, and the consultants were beginning to circle. Tom put the best of them to work providing a series of daily lectures on launch-vehicle design, project management, quality control, simulation software, avionics, guidance, control, and various industry technology surveys. A lot of these consultants were graybeards from the Apollo days who wanted to pass on what had really worked and what had been pretense.

It really wasn't a secret that if you wanted to run a successful project, you needed lots of money, the very best people, and to be left alone. That is, those very best people who were doing the work needed to be left alone. Sometimes this was accomplished, as in the case of the Panama Canal, by giving total authority to a single Army Corps of Engineers general. Sometimes, as with the Fleet Ballistic Missile program, it was accomplished by convincing the U.S. Congress and the U.S. Navy brass that you had invented so many new management tools, such as critical path analysis, systems engineering, and weekly status charts, that they could leave you alone. Sometimes there was somebody like Kelly Johnson who was just too ornery for upper management to interfere with. Sometimes you had a Bill Gates who could move so fast and work so hard that even IBM couldn't keep up with him. Tom and John Forsyth therefore had an understanding. Forsyth would provide the money, Tom would get the very best people, and together they would keep the investors and any one else from getting in the way of the engineering team.

In between reviewing and sorting through all of the launch vehicle data and history that they were collecting, and listening to what the consultants thought could or should be done, Tom had the team doing conceptual design exercises. He purposely started them out with highly improbable approaches—such as a vehicle of all wood structure, a design using only technology available in 1936, or a launch vehicle using nuclear detonations for propulsion. These design exercises put the team through its paces, got them working together, and more importantly, got them used to thinking creatively. They were kept to about two to four weeks' duration and were also used to develop and exercise the design database, CAD modeling system, vehicle trajectory simulations, and the concurrent engineering methodology.

Concurrent engineering is a widely applied approach in which engineers from many disciplines work as a team to simultaneously design a single product. This is best accomplished through the use of an interactive computer database, which is the repository of all of the design details, as they are regularly submitted and updated by the engineers, in the various disciplines. That way, the pieces of the vehicle and system design—performance, weights, propulsion, controllability, aerodynamics, structures, thermal protection, together with ground facilities development, production planning, maintainability engineering, and costing—could all move forward simultaneously.

Tom had also hired some industrial psychologists to develop business/engineering simulations of the vehicle development process. Their work was based on that of Dr. Lia DiBello and her team, who had proved so successful in turning around dysfunctional businesses. As modified for the problem facing Tom, DiBello's techniques were applied to develop simulations of the engineering development, flight test, and production of a launch vehicle. Using these simulations, the engineering, management, and support teams conducted exercises running from two days to a week, in which real strategies for working more effectively as a group were experienced. By this means, Tom hoped to build a team that had the equivalent of many years' of experience working together developing and producing aerospace systems.

CHAPTER 3
Gateses, Jobses, and the Laureates' Lemma

WHILE Tom was putting the engineering team together and conducting his design exercises, Forsyth was conducting what he considered to be a "constitutional convention." He had gathered some of the best and brightest experts from the fields of business, history, and economics, along with experienced executives, managers, and venture capitalists from a broad spectrum of industries. They were meeting during the day and studying at night, trying to really understand what it meant to open a new frontier, and how it could best be accomplished.

On the one hand, space—the "Final Frontier"—was like the New World, except that transportation costs were even higher, and trip times could in some cases be longer than during the days of wooden sailing ships. In spite of the enormous obstacles, the New World had brought immense economic benefits to the Old, and the settlements and colonies there had become economically viable almost immediately. And as had been the case with the New World, the economic value of what was out there on the moon, Mars, and numerous asteroids might very well approach or even exceed the value of the Old World's resources.

But, it was difficult to approach the problem of gaining access to the nearby planets and asteroids as a business proposition. Analysis in this direction had certainly provided the underlying motivation that had brought the Group of Seven together. Merely believing in and being committed to the idea that the space frontier could and should be opened up, however, was not enough to point to an obvious way to overcome the major obstacle: the very high cost of space transportation. The whole transportation segment had to be transformed from a limited-capacity, limited-demand market to one in which costs would begin a downward spiral and demand would really take off.

After a week of picking apart the conquest of the New World, it seemed that perhaps a solution to the problem of how to get the whole thing going would more likely be found by looking at certain new industries and understanding just how they had become viable. Each member of the Group of Seven picked a technically driven industry—the railroads, distributed electrical systems, small electric motors, computers between 1948 and 1978, the personal computer between 1978 and 1995, the computer disk drive systems during the same time period, the Internet between 1993 and 2005—and delved into its history.

On the second day of presentations, Forsyth put forward his belief that the $3 or 4 billion that the Group of Seven had available was clearly inadequate to open up the solar system for exploration and development. Much more would be needed—tens of billions, hundreds of billions, perhaps even trillions. The business models had to show access to that kind of capital, but that didn't mean that AM&M would have to raise all of those tens or hundreds of billions, or even make those billions in profits, at least not in the near term. After all, someone had once estimated that Microsoft only had about 3% of the total market for personal computer software and hardware, but no one could dispute their dominance in that industry!

(My agreement with Forsyth gave me access to all of the Company's meetings, but the Group of Seven balked at seeing their names in print, and so I will just refer to them by numbers.)

At this point, member Number 1 got excited. "This is ridiculous!" he thundered. "I won't have any part of this venture if it is going to be NASA all over again, with countless billions needed to do anything! The whole idea of this group is to make money on low-cost space transportation, and if it can't be done for the money we have, then it just can't be done!"

John Forsyth tried to calm Number 1 down. "First of all, I'm not talking about billions and billions to do just *anything*. I'm talking about what it will take to underwrite the opening of a new frontier across the solar system. Getting that kind of money flowing—into and out of many companies, not just AM&M—will be necessary. And to make that happen, we need to get to a point where a consensus develops, that space is the next 'Big Thing.' A point where people won't bother asking 'how can I make money investing in space,' but rather 'how can I get in on this space thing.' That kind of business climate is what we want, to ensure the ultimate success of AM&M."

"And to make that happen, we need to get to a point where a consensus develops, that space is the next 'Big Thing.'"

Forsyth went on to propound that a review of the development of new markets clearly shows that, at certain points in history, a consensus develops that a new industry is going to arise. But what stimulates the development of that consensus? After all, the basic Internet had been around for 25 years before Marc Andreessen wrote Mosaic and then teamed up with Jim Clark, the former Intel CEO, to create Netscape. What was it that had gotten the Internet going? Clark had just given up on an investment in interactive television. The computer industry moguls, the Bill Gateses of the world, all had thought that high bandwidth was necessary to make money. And they were sure that the product that would be sold on interactive broadband would be television programming, pay-per-view movies, and the like. So, although the Internet existed and people had been using computer bulletin boards to exchange messages, software programs, and yes, even pornography, for years, no one had been interested in investing in low-bandwidth networks other than the government, which used the Internet for exchange of information between universities and government weapons labs. Forsyth asked, "Do you remember when it was that people got really excited about the Internet and everyone became instantly convinced that it was *the* place to invest?"

Number 1 couldn't immediately put his finger on it. Forsyth supplied the answer, "As soon as people *perceived* that money could be made on the Internet. That's when the tremendous interest and hubbub developed around it. Huge sums of money were spent developing websites and companies as the Internet developed. Of course the Internet had gotten overheated and the bubble had collapsed. But, even after the shakeout, the Internet has proven to be a huge economic engine, not only for the United States, but the entire world economy, by substantially reducing the cost of transmitting and retrieving information."

THERE was something inherently human, and perhaps particularly western, about the concept of the next big leap in technology or progress. It was an underlying perception that something new—not simply an incremental change, but a wild revolution—was just around the corner. And whether it was the personal computer or the automobile, it seemed like these revolutions happened almost overnight, even if the basic technology had been in place for some time. After all, the first personal computers differed hardly at all from the systems that Digital Equipment Corporation (DEC) had been selling for years. Even the price differential between DEC's lowest-cost PDP8 minicomputer and the do-it-yourself Alpha computer kit was not large enough to explain the personal computer revolution. It was rather the excitement engendered by the feeling that now real computers were going to be available to everyone. Certainly the founders of Apple, Microsoft, and the myriad other pioneering but long-dead computer companies now consigned to the dustbin of history believed that this technology was going to revolutionize the world. Indeed it had, but not exactly in the way that the PC trailblazers had envisioned. It actually wasn't until the early 1990s, some 20 years after the revolution had begun, that personal computer users started to see some real increase in productivity. But the important thing to note is that the belief in the importance of the technology came before the actual success. If in retrospect it seems obvious that success was ensured, in the early days it was not. Steve Job's prescient observation, after visiting the computer lab at Xerox PARC and seeing the graphical user interface they had developed, was "that it was immediately obvious to anyone that this was the future of computing." Of course, we know it certainly hadn't been obvious to the management of Xerox and wasn't really obvious to business until Bill Gates brought out Windows® 95. Still, it was obvious to people like Jobs. And it was indeed the future of computing.

Forsyth argued that there already existed a widespread belief, if not a consensus, that the space program and space exploration would open up new frontiers and provide a source of vast new wealth. This enthusiasm had been bubbling to the surface for over 100 years. It came in waves, generated by new thinkers, new writers, and new engineering advances, getting stronger each time in spite of numerous setbacks and disappointments. It started with the space pioneers of the turn of the last century—the American Goddard, the Romanian–German Oberth, and the Russian Tsiolkovskii. It was further stimulated by the rocket societies and their experiments in the 1930s; by the burgeoning science fiction literature of the 1930s and 1940s; by the rocket and missile developments that came out of WWII, by Von Braun and his *Colliers* magazine articles of the 1950s, which included Chesley Bonestell's wonderful spacescape paintings; and also by the movie industry, which provided a host of space and science fiction films. Then came the dawn of the Space Age in 1957, and millions of people around the globe went out each evening to see Sputnik soar overhead. With the 1960s came the Space Race between the United States and the Soviet Union. Four years after the first satellite went up, there were men in space. In another eight years, there were men on the moon. By the time of the first moon landing in 1969, there was wild enthusiasm that the opening up of space to everyone was just around the corner.

Then reality set in. The U.S. government, having won the Space Race, decided that there were other more important ways to spend huge amounts of money. The Apollo series was trimmed of the last three lunar expeditions, and plans for moving on to Mars were quietly shelved. Next came the space shuttle—a new type of reusable launch vehicle that wouldn't be discarded every time it flew, and thus would make space access cheaper and more routine, with up to 60 flights per year promised. The bitter disappointment that resulted when it became obvious that the shuttle would never come close to reaching those lofty goals led to disillusionment and cynicism in some circles, and at the same time spurred an interest in alternative private space launch systems. Even as NASA's next big project—the International Space Station—failed again to live up to the promises made for it, there remained for many a stubborn belief that space was going to be important. The many were still waiting for that something, that spark, that would make space the next Big Thing, where real progress would finally be made, and so too would real money be made.

Forsyth had once heard Herbert Kroemer, a Nobel Laureate in physics, expound his favorite proposition or lemma, "The Futility of Predicting Applications," which stated that "the principle applications of any sufficiently new and integrated technology always have been and will continue to be applications created *by* the new technology." Put simply, with any sufficiently new technology, we just don't know what use it will really be until after it is brought into being. Forsyth believed that low-cost space transportation—and he still wasn't sure what low-cost was, perhaps $500 a pound, $200 a pound, $50 a pound, whatever—would lead to more than incremental change. It was going to be one of those innovations that would have a big impact on life as we know it. Just how big, he really couldn't say. And whether the eventual applications included space tourism, which had become popular of late, or solar power satellites, or space colonies, or material processing, or extraction of resources from the asteroids, or something completely undreamed of, the trick was to find a way to bring about low-cost space transportation before you knew what to do with it.

Even if you guessed correctly what it was that would create a huge growth in demand for launch services as costs came down, that demand wouldn't–and couldn't–come into being until the cost did come down.

The last 40 years had seen a lot of futile effort by space enthusiasts to find the one magic product or market that would justify building a truly commercial space industry. Well, it hadn't been found, and frankly, Forsyth was tired of waiting for it. And if you accepted Kroemer's lemma, it couldn't be done that way anyhow. Even if you guessed correctly what it was that would create a huge growth in demand for launch services as costs came down, that demand wouldn't—and couldn't—come into being until the cost did come down. What was needed was the classic "self-reinforcing spiral" that Bill Gates knew so well.

CHAPTER 4
Build Big, Build Many, or Use it Again

AS the Group of Seven adjourned for the day, I trailed along with Forsyth who had decided to go down and talk with Tom about the traditional approaches to low-cost space transportation. We caught him going through a stack of resumes, trying to find those few really good engineers with that rare combination of smarts and the more elusive hands-on engineering excellence. Tom knew he would need the best that he could get.

Forsyth asked Tom to review the three "standard" methods—the big dumb booster approach, the "get the aerospace-military procurement people out of it" approach (sometimes also known as the "refrigerator theory" after the Chinese refrigerator factory, which was also used to mass produce the rocket engines for their launch vehicles), and third, the "make it like an airplane" approach, which was based on a totally reusable vehicle that would be operated like an airplane and therefore would have operating costs like an airplane.

I piped in and asked Tom to elaborate a little bit on each of these approaches for my benefit. Tom began by describing some work done back in the early 1960s by Arthur Schnitt at the Aerospace Corporation. Schnitt had come up with a mathematical theory that attempted to apply to launch vehicles the design criteria commonly used for commercial aircraft, in which the cost of adding a pound of weight to the various parts of the airplane is derived. In an airplane, if you add weight to the landing gear, the structure, the engine, or the wings, this will in turn require additional weight in various other parts, in amounts varying with exactly where the original weight increase occurred. This action will then have an effect on the overall economics of that particular aircraft, again depending on where the weight was added. For example, an additional pound of weight in the wings is more expensive than a pound of weight in the cargo compartment. Although Schnitt's theory was somewhat obtuse, its practical importance lay in the concept that overall launch-vehicle designs should be optimized for cost, not for minimum launch weight, as was standard practice with the early military boosters. The concept was initially called "design to minimum cost." In the launch-vehicle business, it was already a well-understood principle that if the final (upper) stage of a multistage vehicle gained a pound in weight, it lost a pound of useful payload. But if the first stage gained a pound, substantially less payload was lost—sometimes only a tenth of a pound. Schnitt saw that this meant that different cost criteria should be used for different stages, with the upper stages requiring, and actually justifying, higher costs in order to achieve a lighter design.

The vehicle concept that Arthur Schnitt had developed was based on using very simple technologies to build large vehicles that had rather low efficiencies. Whereas the Saturn V placed approximately one pound in orbit for every 22 pounds of vehicle liftoff weight, Schnitt's design placed one pound in orbit for every 40 to 50 pounds of vehicle liftoff weight. Hence the sobriquet "big dumb booster." Even if "big and dumb," it would—at least theoretically—be much cheaper to build and fly.

Schnitt's booster, and virtually all later attempts to build on his concept, employed simple but heavy high-pressure propellant tanks that were filled with fuel or oxidizer, and then pressurized using an inert gas, main tank injection, or by heating the propellants to raise their vapor pressure. The high pressure forced the propellants into the engines. This eliminated the need for costly and troublesome turbopumps to do the forcing. The engines likewise were of simple design, using relatively few parts, and were also relatively low cost to construct. In the 1990s, Beal Aerospace tried such an approach but with lighter-weight composite tanks. Even though they built and successfully test fired an 800,000-pound thrust rocket engine, they withdrew from the field before flying their vehicle, at least in part for political reasons, with a public statement by their CEO to the effect that competing with NASA was simply not possible.

The next approach, using the "refrigerator theory," involved selecting an existing vehicle such as the Delta, or perhaps one of the Russian or Chinese boosters, buying a license to produce it, redesigning it for manufacturability, and then putting it into production using, not the overpriced engineering and manufacturing methodologies and ideologies of the defense contractors, but rather standard industrial techniques such as were employed by the Chinese in their refrigerator factory. To a certain extent, the Chinese and the Russians could be said to have used this approach. Their designs were clearly not low tech, but they were, as near as anyone could tell, a lot cheaper to produce than the comparable U.S., European, or Japanese boosters. The concept was simple: put a launch vehicle into mass production on a world-class, lean manufacturing production line, and see if the cost fell by a factor of 10 or more.

The third approach was intuitively very appealing, if not well supported by industry experience to date. It was based on what everybody knows about airplanes: they are highly sophisticated machines built by well-established and profitable aerospace companies, with levels of reliability and efficiency that make air transportation the safest and most economical means of transporting people over long distances. And although the aircraft themselves are not cheap—prices for commercial airliners range up into the hundreds of millions—they are at least affordable enough to be operated by private airlines and owners the world over. If you could do that with something as sophisticated as a Boeing 747, why couldn't you build a rocket that did the same? Because this approach was based largely on "gut feel," it was sometimes taken to the logical extreme that it would most likely be successful if the reusable spacecraft looked like an airplane, with wings and wheels—if it looked like an airplane, it should cost like an airplane, the reasoning went.

THE monumental failure of the space shuttle to achieve not only its original performance targets, but also any cost reduction at all over the expendable vehicles that it was supposed to replace, had not dampened the widespread enthusiasm for this approach. Since the mid-1980s, NASA had been funding one airplane-look-alike design study after another and was even now preparing to let final bids on their latest concept, boldly proclaiming that this time it really would replace the shuttle and dramatically lower the cost of space transportation.

Forsyth asked Tom where he thought the solution lay. Tom leaned back, thought for a bit and said, "I think it's in the fully reusable vehicle that functions like an airplane, although I don't think it will look like an airplane." He went on to elaborate, however, that the problem of analyzing the feasibility of such a vehicle is much more difficult, compared to the problem of analyzing a big dumb booster design or a concept for an efficiently mass-produced existing booster.

"I think it's in the fully reusable vehicle that functions like an airplane, although I don't think it will look like an airplane."

"Well," said Forsyth, "why don't you put together three design teams—one to come up with a big dumb booster-type vehicle, and take a look at what we could do to reduce space transportation costs with that. Put a second team together—ignoring for the moment the cost of a license for the technology—and come up with a good estimate of how much costs could be reduced by using commercial best practices manufacturing. Finally, without getting bogged down in endless design trades and architecture studies, try to pin down the general parameters for the economics of a fully reusable, airplane-like vehicle, again without wringing out all of the particular technical issues and solutions—at least for the time being. With that kind of information in hand, maybe we'll be able to see more clearly how a market could be created for whatever looks to be the most feasible approach."

Tom allowed as how he thought they could get something useful together in three or four weeks that would lay out the general parameters for the different approaches. He got to work right away, forming from the ranks of his ever-expanding staff three teams, one for each of the approaches.

In performing a vehicle concept study, it was important to use parametric values for such things as weight, cost, and performance estimates in order to save time. Besides, the team was not going to be designing the complete vehicle anyway. It was also very important to use numbers that came from systems that had actually been built, in order to keep the designs honest. In Tom's experience, the wildest flights of fancy—improbable or often impossible designs—came from engineering teams that got too far away from their own experience. They would leave behind the realm of hardware that had actually been found to work, and even what had already been found *not* to work, and start making it up as they went along. Or, just as bad, they started using other flights of fancy as their starting point. No doubt progress is at times made by geniuses who leapfrog over the conventional wisdom, but genius is where you find it. It's not usually the prime mover in a successful design program; engineers are most likely to succeed if they incrementally build on known technique. After all, the graphical user interface was created by a genius at Xerox, but it was Steve Jobs who built something useful out of that creation.

BACK in 1996, a man named John Whitehead wrote a paper (AIAA Paper 96-3108) in which he showed that the weights of typical launch vehicle stages had not changed substantially over time. If tank weights have not improved in 50 years, it's best not to assume you can dramatically beat history. Whitehead also noted that tank weight as a function of propellant volume was essentially constant over a wide range of tank sizes, from a few tens of thousands of pounds to well over a million pounds. He highlighted the fact that the weight of a tank capable of holding a given quantity of propellant was directly proportional to the propellant density because the tank weight was almost completely dependent on the volume of the propellants.

If tank weights have not improved in 50 years, it's best not to assume you can dramatically beat history.

Whitehead also addressed the other major weight contributor in a typical stage, the rocket engines. For dense propellant combinations such as liquid oxygen (LOX)/kerosene, the engine typically weighs 1% of the thrust (for a thrust-to-weight ratio or T/W of 100); for lower density propellants such as LOX/LH_2, the figure is in the neighborhood of 2% (T/W of 50). A third but smaller contributor is the propellant pressurization system weight. This system provides the gas pressure that is used to force the propellants through the piping of the feed system and into the turbopumps, which feed the engines. Finally, there is the weight of the residual propellants that remain in the tanks or the feed lines and are unused by the engines.

The remaining structural elements in the stage, the engine thrust structure, which transmits the thrust of the engines to the loaded propellant tanks, and the payload or upper-stage support structures were not considered by Whitehead because these tend to be much more design-specific. The purpose of his paper had been simply to point out which structural elements of a typical launch vehicle stage were more or less mature in terms of achievable weights, and which were likely to benefit from better engineering in order to reduce their weights. Whitehead's main concern was the problem of single-stage vehicles. For a single-stage vehicle, the rocket equation $MR = e^{(\text{Velocity}/Isp \times 32.2)}$ tells you, for a given engine *Isp* (i.e., jet exhaust speed divided by acceleration caused by gravity) and a total desired velocity change, what fraction of the initial stage mass would achieve that velocity change.

The mass ratio (MR) is the ratio of the initial mass of the stage to the final mass after all of the propellants necessary to attain your desired velocity change have been burned. If that desired velocity change is equal to that required to go from zero at your launch point to orbital velocity, then it will tell you how much of that initial mass will reach orbit. Whitehead, by characterizing tank, engine, and propellant feed system weights as percentages of the total final or empty weight of the stage, provided a fairly accurate means of estimating how much weight could be allocated to the other structural and vehicle components and, more importantly, how much payload could be included. In other words, if you know the percentage of the final weight that must be allocated to the tanks, engine, and pressurization system, then you can determine how much is left for the other vehicle systems and the payload.

Considering the parametrics of a pressure-fed design, three factors must be considered that contribute to lower performance as compared to a turbopump-fed stage. The first is the greater weight of tanks strong enough to support the higher internal pressure needed to force the propellants all the way to the combustion chamber. Second, because this pressure is typically 5–10 times higher than the internal pressure required for a pump-fed stage, the weight of the gas used to pressurize the tank is also proportionately greater. Lastly, the engine specific impulse (*Isp*) is also typically lower for a pressure-fed stage because the engine chamber pressure tends to optimize at a lower value when the increased weight of the pressurization system and tanks is considered. The higher the tank pressure is, the greater the weight of the tank and the pressurization system, which drives the design toward a low-pressure engine. Thus by eliminating the turbopumps, which are complex and expensive both to develop and to produce, the designer accepts higher weights and generally lower specific impulse.

Because of the high tank weight, pressure-fed vehicles have typically utilized dense propellants and lightweight pressurizing gases. For a typical pump-fed stage the tanks weigh approximately 1% of the weight of the propellants (for propellants with a density about the same as water). If the pressure in the tanks is increased by a factor of 10, one would expect the tanks to grow to about 10% of the weight of the propellants. To minimize this weight growth, high strength-to-weight materials are typically employed.

It would seem logical, therefore, to use composites with their very high strength-to-weight ratios. However, for a stage with a propellant capacity of a million pounds (again, assuming the density approximates that of water) a tank 16 feet in diameter would be 80 feet long. This presents some serious manufacturing difficulties in laying up and curing the tanks; the highest strength composites require autoclave curing—and that would take a very large autoclave. In the past, high-strength steels like maraging steel, which has very good strength at cryogenic temperatures, or high-strength, low-alloy steels such as HY-180 have been suggested. Both of these materials have been used for solid rocket motor cases.

In many ways, a pressure-fed vehicle is similar to a solid rocket. In a solid rocket all of the propellants are contained in the combustion chamber where, during operation, the pressure is typically in the range of 500–1000 psi. To withstand these high pressures, composite-wound cases have been used in many missiles and rocket boosters. However, for the extremely high-performance upper stages, such as the Thiokol Star 48 motor, the case is a thin-walled titanium shell weighing only about 3% of the weight of the propellant contained. The solid propellants used in a Star 48 actually have a density about 1.6 times that of water. Thus, it is theoretically possible to build a high-performance, thin-walled titanium tank that weighs approximately 5% of the weight of its contents, if those contents have a density about the same as water.

THE concept that the first team came up with was a two-stage launch vehicle. Considering the low performance of pressure-feds, three stages might provide higher performance, but Tom reasoned that, for the purposes of estimating costs, two stages would be close enough. The costs for a vehicle with two stages were not likely to be substantially higher than those for a three-stage vehicle because the higher performance of a three-stage vehicle would have to be weighed against the cost of the additional stage.

The first stage had a gross weight of 900,000 pounds and was constructed of high-strength rolled AerMet 100 with a yield strength of 261 ksi, an ultimate strength of 300 ksi, and an elongation of 10% before rupture. AerMet 100 is a recent development, a new martensitic superalloy that is stronger and less expensive than other superalloys such as Inconel 718. HY180 was also a possibility; it is a high-strength low-alloy steel with good ductility and good weldability. It had been proposed for use in the original big dumb booster studies. The first stage was actually one pressure vessel divided into two tanks by a common bulkhead. A common bulkhead—the dividing wall between the oxidizer and the fuel tanks—could be constructed from thinner material because it would be subjected to lower stresses than the other portions of the tanks; it would only need to withstand the pressure differential between the two propellants.

An interesting property of materials, which pressure vessel designers always have to keep in mind, is that it is important not only to use a strong material, but also a material that resists the formation of cracks. After all, glass is as strong as steel, but nobody would like to sit next to a glass tank with 3000 psi compressed gas inside. If such a tank started to crack, you would be sitting next to a bomb. For metal tanks, it turns out that the material is toughest—that is, the most resistant to tearing—at a particular thickness. Very thin materials are very easily torn; aluminum foil, for example, is usually torn by hand when you need a piece. An aluminum pop can is rather less prone to tearing, and a 1/8th-inch-thick aluminum plate much less so, and not just in proportion to the greater thickness. After all, even the soft unalloyed aluminum used for foil has a yield strength of about 10,000 psi, and with a thickness of two or three thousandths of an inch it would seem that a force of 30 or 40 pounds

After all, glass is as strong as steel, but nobody would like to sit next to a glass tank with 3000 psi compressed gas inside.

would be necessary to press your finger through a sheet of it. But, as we have all experienced, it is quite easy to tear once you get it started!

The upper stage would weigh approximately 100,000 pounds and consist of an aluminum vessel overwrapped with S-glass. The majority of the aluminum was subsequently removed by dissolving it with sodium hydroxide; this simplified the tooling and also the plumbing of the vessel.

The propellants selected for both stages were NH_3 and N_2O_4. Pressurization was to be achieved by heating the propellants to 145 and 232°F, respectively, an approach known as VaPak that had been developed by Aerojet in the early 1960s. Burt Rutan had used vapor pressurization in the N_2O oxidizer tank for the hybrid engine of his SpaceShipOne. Using VaPak, the propellants would be raised to an initial pressure of 400 psi. As the tanks were emptied, the pressure would drop, allowing the hot propellants to boil and thus provide more pressurant gas. As propellants are expelled from pressurized tanks, their temperature will drop so that by the time all of the liquids are drained from the tanks, pressure has fallen 20–30% from its original value. Approximately 7–10% of the propellants would be left in the vapor state after all of the liquids were expelled.

Vapor pressurization has some disadvantages: heating the propellants reduces their densities, and in order to achieve low propellant residuals (the left-over, unusable propellants) the engine must be operated for a time on the propellant vapors. However VaPak has many advantages. Residuals are low because, if one propellant's vapor pressure decays more rapidly than the other, the engine can still burn them; the mismatch in the relative amounts of each propellant results merely in a change in mixture ratio, rather than incomplete utilization of the propellants. Mixture-ratio changes—although they do affect *Isp*—do not affect performance to the same extent that excessive residual propellants do. The vapor pressurization technique has the advantage of being completely automatic, that is, no valves, regulators, or heat exchangers are required for it to work. In addition, the liquid propellants will tend to vaporize as they experience pressure drops while flowing through the feed lines or crossing the injector. This vaporization of the propellants improves mixing and results in high efficiency with a simpler, less complicated injector design.

Finally, because every drop in the pressure of the propellants causes additional vaporization, the propellant feed lines always contain some gas, which makes the propellants at all times compressible. Many rocket engines, especially in pressure-fed stages, have been plagued by combustion instabilities caused by an interaction between pressure oscillations in the thrust chamber; this is the well-known "pogo effect." A pressure oscillation in the engine increases the chamber pressure, which inhibits propellant flow into the engine, which in turn causes a decrease in chamber pressure, which in turn allows more propellants to flow into the engine, thus generating an even greater pressure pulse. The result is severe pogo-like vibrations that can destroy the vehicle. Some of the research done to eliminate pogo in the Saturn F1 engine showed it could be overcome by injecting a small amount of gaseous helium into the propellants (so as to make them compressible). The oscillations in the engine had relatively little effect on the flow of the now-compressible propellants; pressure waves are not as easily transmitted by compressible fluids.

With the VaPak approach, a tank pressure of 400 psi could be used with a chamber pressure of 200 psi. Moreover an *Isp* of 215 seconds at sea level and 250 in vacuum appeared to be achievable with NH_3 and N_2O_4. A 16-foot diameter tank with a nominal design pressure of 400 psi resulted in a tank wall thickness of 0.15 inches, which was thin enough to provide reasonable material toughness. The tanks could be designed with a safety factor of essentially 1.0; by using simple heating of the propellants for pressurization, you avoid the possibility of transitory overpressure in the tanks. Any increase in pressure would immediately cause some of the gas to condense, thus providing self-limiting pressure regulation.

First-stage tank weight was approximately 10% of that stage's total weight or 90,000 pounds. With a vacuum *Isp* of 250 and an upper-stage weight of 100,000 pounds, the first stage produced a vacuum equivalent delta-V (change in velocity) of 13,400 fps. The upper-stage tanks would weigh about 3600 pounds, or between 3 and 4% of the weight of the fully loaded stage; total structure weight was approximately 6% or 6000 pounds. The upper stage, with a slightly higher *Isp* (280) and somewhat lower mass fraction, could place 10,000 pounds into low Earth orbit. The lower-stage structure was estimated to cost $25 per pound and the upper-stage structure $250 per pound. The cost of the propellants, NH_3 and N_2O_4, would be about $0.50 per pound of propellant or $50 per pound of payload, giving a total cost for the vehicle of $425 per pound of payload.

TEAM 2 had deviated somewhat from the initial approach of taking an existing launch vehicle and reengineering it for commercial-type mass production. They ended up designing a single-stage expendable vehicle based on the Japanese LE-7 engine. This pump-fed engine, using LOX and LH_2 at a mixture ratio of 6:1, has a sea-level thrust of 250,000 pounds, a vacuum *Isp* of 440 seconds, and a very respectable thrust-to-weight ratio of nearly 65. This engine was flight proven, having been used in the H-2A launch vehicle.

Team 2's vehicle would not have quite the 10,000-pound payload capability that was the design target for each team. However, because it used an existing engine with good performance, Tom had allowed them to go ahead with this concept. Team 2 calculated that 12% of the weight of the vehicle would be injected into orbit. With reasonably efficient thrust structure and payload support system designs, this meant they could get approximately 3% of the launch weight for payload; for a 200,000-lb liftoff weight that meant 6000 pounds of payload.

Quick inquiries with the Japanese—although going somewhat beyond the ground rules for this study—had indicated that they were quite willing to license the technology. Costing the LE-7 in mass-production mode, using comparisons with the highest production-rate jet engines, indicated that a rate of approximately 25 engines per day could bring the cost of the engine down to $250 per pound, or approximately $1 million per engine. The cost of the engine divided by the payload meant the engine contributed $166 per pound to the cost of launch. The lightweight aluminum tanks, at high production rates, were estimated to cost in the neighborhood of $20 per pound. The vehicle would have only about 3 pounds of structure per pound of payload, and so the cost of the structure of would be approximately $60 per pound placed into orbit. Thirty pounds of propellant per pound of payload, at a cost of $.25 per pound would add $7.50 per pound of payload. Total cost should thus be about $250 per pound. Twenty-five vehicles per day would cost about $37,500,000 per day. Development cost seemed within the $3 billion capital budget. Thus, it appeared feasible to mass produce a pump-fed, single-stage expendable vehicle with a cost of less than $250 per pound of payload to orbit.

THE third approach—a fully reusable launch vehicle—was necessarily limited to parametric studies, given the time available for this design exercise and cost estimation.

As Len Cormier of the Third Millennium Aerospace Company had argued persuasively some two decades earlier, the most important cost parameter for a reusable launch vehicle is its development cost. The cost of propellants for any reasonable two-stage launch vehicle, which should be able to place 1–2% of its launch weight into orbit as payload—for a wide range of commodity-type propellants such as liquid oxygen, hydrogen, kerosene, liquid propane, or liquid methane—is in the $5–10 per pound range. Plugging in a maximum development cost of $3 billion, an additional $2–5 billion for production startup, and providing for the necessity of producing over 100 vehicles to get down the learning curve, so that for the first 100 vehicles the average production cost was approximately $200 million, the total cost of vehicle production would be about $28 billion for the first 100 vehicles. Assuming a payload capability on the low side at 5000 pounds, and a targeted goal of $500 per pound, a total of 56 million pounds of payload (representing 11,200 flights) would then be required before the total payload weight launched times the cost per pound would equal the development and production costs. Obviously, when operating costs and the cost of money were added to this calculation, tens of thousands of flights would be necessary before the fully reusable vehicle would be economical. This was not really surprising, considering that for aircraft such as the Boeing 737 and 727, the number of flights per year for all airframes of a type exceeded one million.

The results from the three studies were taken to the Group of Seven. After the presentations, the members certainly felt more educated and enlightened about the various approaches and were genuinely encouraged that all of the approaches were shown to be workable options in pursuit of the low-cost spaceflight objective. However, Forsyth proceeded to throw some cold water on their enthusiasm.

"This is nonsense," he said. "Here we are talking about tens of flights per day, when the worldwide market for launch vehicles in this size range is much less than a hundred per year, and two-thirds of those payloads wouldn't look at our vehicle even if it were free! I may have a billion dollars to spend, and between us we could spend three to five billion, but that's clearly not enough to build any of these systems if it's going to take $5–10 billion dollars per year in launches to achieve our cost goal. There aren't nearly enough customers with enough payloads to come close to paying us for all that launch capacity. Now it might be that if some government wanted

"Here we are talking about tens of flights per day, when the worldwide market for launch vehicles in this size range is much less than a hundred per year, and two-thirds of those payloads wouldn't look at our vehicle even if it were free!"

to mass produce a vehicle, spending $5–10 billion per year, eventually some new uses, some new demand, might arise for low-cost space launch. But gentlemen, we are not a government and, frankly, I don't think that building and throwing away 10, 20, or 30 launch vehicles a day is ever going to produce the paradigm shift we want. There must be another approach. There must be some way we can start small—our three billion *is* small compared to these numbers—and make something that will allow us to grow the market for low-cost space transportation. We have to actually make some money, not just burn it! We'll *have* to make money if we hope to get access to the capital market to fund the full-blown opening of the frontier."

After the engineers were sent back to the shop for more design exercises, Forsyth and the rest of the Seven batted the problem around. Number 3 said, "Well, it's clear to me that we can't make money selling launch services. The worldwide market is less than a few billion. If we drop the price by a factor of ten, then that market is less than a couple hundred million. Besides, most payloads go on national launchers, and customers are more concerned about track records and vehicle environments and large unitary payloads. If we offered services for $500 a pound, I doubt whether we'd get even $50 million a year to start with. Sure, if you offered it at that price, demand might grow rapidly and sales would go up. But at $50 million a year in revenue, we'd long be broke before there was a big enough market to support the kind of investment which you and I now know is necessary to build a useful launch system."

Number 5 piped up and said, "Let's look at airplanes. Most of the early airlines lost money most of the time, even with government subsidies for the mail routes. Most of the airplane manufacturers lost money, too, but not all. Boeing did okay, even after it was forced to sell its airline."

Forsyth picked up on that idea and said, "Yeah, what about selling the launch vehicles rather than the launch services?"

Number 6 objected. "That doesn't make any sense! What are we going to sell? Even for the low-cost mass-produced single-stage or the two-stage pressure-fed, the necessary infrastructure alone is going to make them almost as expensive as current vehicles for low launch rates. Besides, nobody is going to pay us $1–2 billion a year for a bunch of vehicles they don't have any use for. There's just no market there!"

At that, Number 3 responded, "I think it would have to be the reusable vehicle, operated just like an airplane—it has got to be like that."

But Forsyth asked, "Sure, but even so, why would anybody want to buy them if they haven't got a use for them?"

Number 7 jumped in. "But wait a minute! Let's assume for a moment that people would be willing to buy such a launch vehicle. It's reusable, it's relatively inexpensive, costs maybe a quarter billion dollars—no more than a 747. Now if we had a vehicle like that which sold for $250 million, we would only need to sell something like eight per year to get $2 billion in revenue which, as we have discussed, is the minimum to create a viable business."

"Sure, but who are you going to sell eight vehicles to?" demanded Number 5.

Number 3 said, "Well, maybe there'll be enough crazy people out there to just buy them. I mean, some of us must have been crazy when we bought into building the fiber optic backbone. Man, we spent well over $100 billion on that, and we all lost our shirts! It took fifteen years before the demand finally caught up with capacity."

"True indeed," said Forsyth, "it did finally reach capacity. All that fiber optic cabling eventually was used, and as new technologies became available along with it, the whole business turned out to be a lot bigger than we ever dreamed it would be, back at the beginning." Forsyth paused for a few moments, and then said, "Now, people, I think we're on to something, I think the right approach, the only approach, is to build a launch vehicle which we can convince people to buy outright. Then our problem is reduced to selling eight vehicles a year to start with and growing sales in a simple arithmetic progression—10 the next year, 12 the year after, and so on. *That's* a viable business. So, what would launch costs look like then? Number 5, you're the economist. Say we find 8 to 15 somethings or somebodies to buy launch vehicles from us every year—what are they going to do with them?"

"I think the right approach, the only approach, is to build a launch vehicle which we can convince people to buy outright."

"Well," Number 5 replied, "I suppose they'd fly them, but with that much capacity, it would have to be at a loss."

Forsyth said, "That actually sounds good to me! Operating at a loss means that space transportation will be cheaper than it really costs, which should encourage more uses and greater demand." There was some additional murmuring as the group thought about this idea. Forsyth eventually said, "I think we should call it a night. Let's go back and think this over a bit, and tomorrow we can talk about how we might be able to sell some sort of vehicle—what kinds of reasons we could give people or institutions to buy a vehicle. Then we'll get the engineers in here again, and see if they can build something that fits the bill."

CHAPTER 5
Small Market, Small Payload, But Not a Toy

THE Group of Seven reconvened the next day, now focused on identifying what exactly it was that they were going to be selling. As Number 3 had suggested during yesterday's deliberations, the starting point would be an airplane. Not that their space transport vehicle would actually *be* an airplane, or even look like an airplane, but it had to "cost" like an airplane—both to purchase and to operate—and it would need to be pretty reliable, at least reliable enough that loss of vehicles contributed little to the cost of operation. To kick things off, Number 5, the economist, summarized the worldwide market for aerospace vehicles: "Assuming we stick with the $250 million price tag that we kicked around yesterday and a goal of selling about $2 billion worth of vehicles per year, then that dollar value represents only a couple of percent of the world-wide market for commercial and military aerospace vehicles." He continued, "Of course there is not currently any market for reusable launch vehicles, but if we can create such a market, it would be small enough that the money for our vehicles could come from reallocation of resources now going to other aerospace systems." So on a top level, it was at least conceivable that the money was already there to support a market for the proposed vehicle.

Forsyth then sketched out what he thought were or should be the criteria for a sellable vehicle. "First, the vehicle will have to have low operation costs, or at least be able to achieve that after getting down the learning curve. After all, the goal is not to create some expensive new military toy which can only be sold in small quantities like the B-2 Stealth bomber. Rather, it is to sell something that will set off a spiraling decrease in the cost of space transportation. That means lots of vehicles." Next was size—the smaller the better, according to Forsyth. "Even with a very small payload, the introduction of this vehicle will result in a huge surge in launch capacity, if they are truly operated on a routine basis—starting, say, at once a week and pushing toward once a day." In his own mind, he liked a payload of 5000 pounds—large enough to handle about half of current satellites, but more importantly, a significant "pay"-load in commercial terms. He also did not want to aim too low, lest vehicle performance and weight growth problems drive the payload so low that launch costs on a per pound basis could not fall to the very low values he envisioned. Five thousand pounds at $200 per pound was $1 million a flight. At $50 per pound, it was $250,000 per flight, which was down in the range of the more expensive commercial airplanes.

"The vehicle will also have to be manned, he declared."

Forsyth paused, and then continued with an item he knew would be controversial. "The vehicle will also have to be manned," he declared.

Number 5 nearly shouted, "But that will make it prohibitively expensive! Manned systems always cost way more—just look at the shuttle. Its costs are at least several times the going rate for any commercial expendable launcher!"

Unruffled, Forsyth replied, "If it is going to be a reusable vehicle, then it is going to have to be extremely reliable, or it will not be cost-effective to start with. Putting a pilot on board will increase reliability. Ask anyone who has studied the accident or loss rates for unmanned vehicles. They are at least 10 times those for any manned vehicle, which makes them impractical for any other than military or very esoteric civilian uses where the higher loss rates are considered acceptable. Secondly, this thing is going to have some sizzle! Buying a manned vehicle will mean more than just buying a way to transport payloads to orbit; it will mean that whoever buys it—a nation or any other entity—is entering a very exclusive club. A club whose members have paid much to be there: the United States, Russia—and they're barely hanging on—and the Chinese, who have just managed to get in.

"And don't forget the space tourism market that everybody keeps talking about," he continued. "The idea here is that people will be buying the vehicles on speculation—hoping to be part of a new industry. We have to give them as many reasons as we can to justify their buying this capability without a definite plan for using it. And I'm convinced that the prestige and excitement of being able to launch people into space will be critical. Of course," added Forsyth, "I am certainly willing to concede that every attempt should be made to include a safe and reliable escape system for the crew. I also think that the engineers and technicians who build this thing will be more motivated to build in reliability and safety when they know lives will depend on it."

Forsyth ticked off the rest of his criteria. "The vehicle should be completely reusable, not only as a selling point, but also because any significant expendable component will make it difficult to bring the costs down. The idea is to design a vehicle that will provide the opportunity for a real learning curve on the whole system. Also, it will be better to avoid the operational limitations that would come with a vehicle that drops pieces off every time it flies. Finally, the vehicle should be able to be launched from and recovered at a single launch site—not that there will be only one single launch site, of course, but that a customer will only need one, not two or more for launch, recovery, contingencies, whatever. And, preferably, the launch facilities will be simple enough that their cost will be significantly less than the vehicle price, perhaps $100 to $150 million."

After Forsyth put forward his basic requirements for the vehicle, the debate raged on for some time over whether or not there actually existed a market for such a thing. But eventually the realization emerged, and none could really disagree that if such a thing could be built—a fully reusable launch vehicle that would place 5000 pounds and one or more persons into orbit, with a pricetag in the neighborhood of $250 million—reasons would be found to buy them. Consensus also developed that an operational cost that could fall to less than $1 million per flight with reasonable flight rates was critical to long-term success.

Tom was brought back in to answer some technical questions. One of the most interesting points that he made was that 5000 pounds into orbit would require somewhere between 100,000 and 400,000 pounds of propellant. Even the most expensive propellant combination, LH_2 and LOX, costs less than $0.30 per pound. The cost of propellants was therefore likely to be less than 10%—and probably less than 5%—of the target $1 million per flight cost. Other similar transportation systems typically have overall operating costs that are a small multiple (two to three for the airline industry) of the cost of fuel.

The Group of Seven adjourned for a month, to give them each some time to think about the marketing of such a vehicle and its feasibility, and also to give Forsyth a chance to get together with the engineering team to see what this vehicle might look like. Forsyth was not himself an engineer by training, but he had kept abreast of the literature, and he was familiar with all of the ongoing debates about whether single stage was possible, two stage was best, airbreathing was cheaper, and many others. He had a definite opinion about which way he thought made the most sense—ballistic single stage. That is, a vehicle consisting of only one stage, without wings, that flew to orbit and returned to the launch site for a rocket-powered landing like the DC-X, the Phoenix, or the German Beta vehicles. The other members of the Group of Seven, being long-time space enthusiasts, even if not experts in the field of rocket engineering, shared a lively interest in vehicle design. Most had their own strong opinion of what the vehicle should look like, now that the basic requirements for size, cost, and marketing approach had been decided.

CHAPTER 6
Myths, Mistrust, and Trust

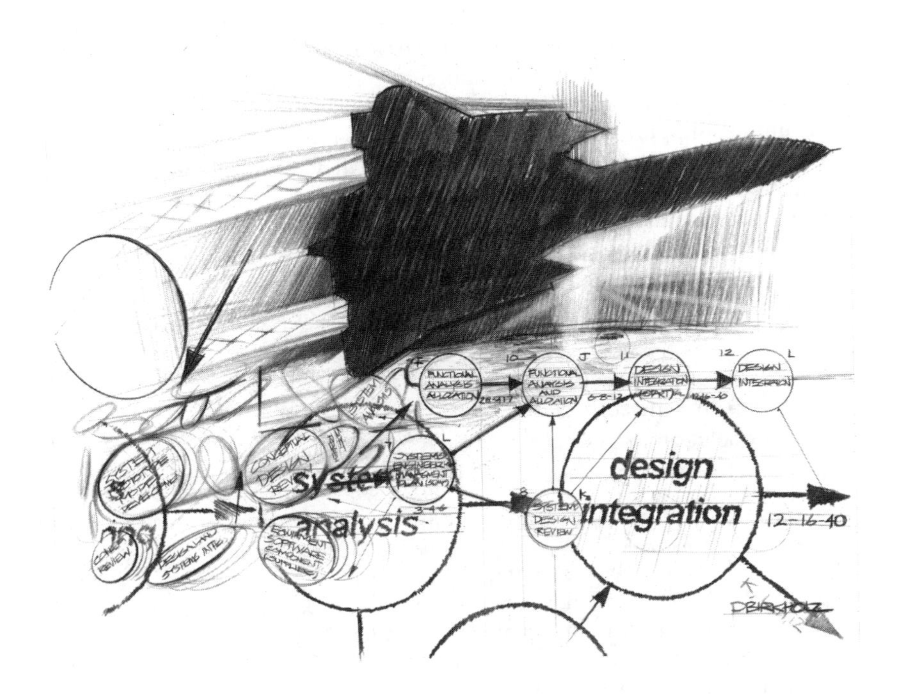

ALTHOUGH John Forsyth had his opinion about what the fully reusable vehicle should look like, he knew he wasn't the one who was going to build it. He had hired Tom to do that, and he knew from reading the history of many aerospace engineering projects that the man who was going to build the vehicle—if he was the right man for the job—would also need to be the one to decide what to build. Forsyth's job was to lay out what he wanted it to do, and then let the engineers tell him what it would look like. He'd been hanging around with old rocket hands for years; he'd read the history of the space program, he'd read *Augustine's Laws*, and he had read the history of the famed Lockheed Skunk Works. Truly innovative systems tended to flow from a single individual—an Admiral Rickover, a Kelly Johnson, a Wernher Von Braun—who was in complete control of the project and who had an integrated vision of what was to be accomplished. So he was quite aware that if there were to be any chance of success, he was going to have to ride the single horse that was Tom and his team of engineers, putting his faith in their understanding and dedication to the task set before them.

Forsyth had been observing his chief engineer for over a year now, and he knew him to be a man possessed of encyclopedic knowledge of launch-vehicle systems, subsystems, design details, and operations. His confidence in Tom had been growing as he watched the engineering team come together under his leadership. He was impressed by the results he had seen come out of the design exercises, and he was especially impressed with the organization he had built, and how he ran it. Tom managed by wandering around, talking to each engineer, and getting them the resources they needed to do their jobs. He had also built up an administrative infrastructure that, although quite lean, functioned effectively to serve and support the engineers. In the major aerospace companies, hapless engineers too often found themselves more or less subservient to those support functions—contract administration, procurement, quality control, human resources, computer support—that were supposed to serve the engineers. Tom treated his engineers like they were the top dogs because they were. The engineers, not the administrators, were the ones who would make this project a success. As a result, they felt valued and supported, and they were working hard. Forsyth was sure that they would "pour on the coals" when they had real meat to chew.

Yes, Forsyth was quite happy with his choice and was prepared to ride that horse into battle. When it came to choosing the design, now that the basic approach had been set by the Group of Seven, it wasn't going to be a matter of endless design studies, but a matter of taking what the man he'd picked to do the job thought was the most suitable approach. Many people—engineers included—seem to think that the correct engineering solution to a given problem is a matter of fact, not opinion, because the underlying constraints on the problem and its solution are physical laws of nature. But Forsyth had enough experience with engineers to know that any design was almost always a matter of opinion. There is more than one way to skin a cat, and which approach would in the end prove to be the most elegant, cost effective, and ultimately successful in the marketplace was often clear only in hindsight. Thus, John Forsyth was firm in his conviction that this was one project that was going to have to be executed in a manner vastly different from typical aerospace development programs.

Many people—engineers included—seem to think that the correct engineering solution to a given problem is a matter of fact, not opinion.

The recent history of those programs was, to say the least, dismal. Fewer and fewer new vehicle programs made it past the "paper study" stage, and fewer and fewer of those that did actually made it into production. Certainly none of the space shuttle "replacement" vehicles had. Forsyth remembered reading an article by Willis Hawkins in the October 1982 issue of *Aerospace America*. Hawkins, who had held a number of senior management positions at Lockheed Missile and Space Systems during the 1960s and 1970s, pointed out that the military services often canceled a development program just before it was ready to go into production because the program was over budget, over schedule, and projecting unit costs way above design goals. So they canceled the tank, the helicopter, the airplane, or the ship and started again from scratch, this time seeking a less capable, but ostensibly lower-cost system. After many more years of design and development, the new system was ready for production. Inevitably, that system ended up costing more—with less functionality and lower performance—than the system which had been canceled. In the meantime, enough money had been spent on research and development for the new system to pay for a sizable deployment of the old one!

THE truth is that engineering is not such an exact discipline that any complex machine, when finally built, will necessarily have the functionality, or the costs, originally envisioned. The trick is to assess what you have, and, if it has value, go with it. Aviation history is replete with stories of success being snatched from the jaws of failure. The P-38 Lightning of World War II fame was a classic example. Designed as a high-altitude escort fighter, it turned out that it didn't perform well enough at high altitudes. Rather than being canceled, it ended the war with an outstanding reputation as a low-altitude fighter, reconnaissance aircraft, night fighter, and attack aircraft—none of which it had been designed to perform. Some 10,000 of them were produced during World War II.

Indeed it has not been uncommon, as conditions and needs change, for airplanes or systems developed for one purpose to go on to other applications at which they excel. The fabled Douglas DC-3 was originally designed as a passenger-sleeper for long, overnight transcontinental flights. It ended up as a short-haul passenger and cargo plane, still going strong in many parts of the world 70 years after it was designed. The SR-71 Blackbird, originally conceived as the A-12 for high-altitude overflight missions, was later modified to use side-looking radar and advanced imaging cameras that did not require overflight, when such flights became politically untenable. And the first LOX/LH_2 high-performance rocket engine, the Pratt & Whitney RL10, actually started life as a hydrogen pump for a hydrogen-fueled aircraft. That aircraft turned out to be such a bad idea that Kelly Johnson, whose Skunk Works had been selected to develop it, canceled the program on his own and sent the money back to the Air Force.

These days, most aerospace programs start with an elaborate set of official requirements. If you try to track down where the requirements come from—tracing them back from the request for proposal, to the program office, to the armed services, to the Pentagon, to the committees of Congress, even back to the cocktail parties of Washington—the trail often ends cold. Nobody really knows from where the requirements come. This means that they are not generally part of any cogent vision and thus tend to shift and mutate from year to year along with the ever-shifting political climate. But these requirements, not the insight of competent engineering managers, drive the system development process.

But these requirements, not the insight of competent engineering managers, drive the system development process.

In the usual NASA or Air Force approach, these requirements are fed into a system development process that generates several hundred (yes, hundred!) possible vehicle concepts. These concepts are then meticulously compared, traded, and downselected to the absolutely best solution. Given the constantly moving design requirements target, one usually ends up asking the question "best possible solution to what?" The whole process is, unfortunately, based on a well-meaning ruse.

Tom, who had worked for a major aerospace contractor in the early 1980s, had personal knowledge of where the trouble had begun. In the 1950s the company he was working for had been faced with a challenging ballistic missile project. The company program managers and government contract administrators had realized that one of the most difficult hurdles they faced was how to get the government agency overseeing the project and Congress to leave them alone. They knew that a successful program stood on three legs: a lot of money, the best people, and freedom from interference by those who didn't understand the engineering and who weren't actually doing the work. With a technically challenging program like this, freedom of action was essential. So the program managers came up with a carefully crafted set of tools to convince the oversight agency and Congress that they didn't need to worry about whether they (the contractors) knew what they were doing.

THIS took the form of new management techniques that were used to show that they were using such advanced, innovative, and scientific tools that they were beyond questioning. The main components were systems engineering, program evaluation and review technique (PERT) scheduling, and status review charts.

Systems engineering was an outgrowth of operations research, which had contributed so much to winning the Battle of the Atlantic against German U-boats during World War II. Operations research was used to apply mathematical techniques to strategic or tactical battlefield problems and later to business management problems. These techniques gave new insight into such problems and provided analytical means of solution. Systems engineering sprang from the idea that, as engineering projects became larger and more complicated involving many diverse disciplines that all had to work together, a similar scientific approach was required. To implement this new method, a new class of superengineers was created—the systems engineers. These men wouldn't actually be doing any designing; they merely had to understand the entire project, all of the technical disciplines involved, and oversee the project to ensure that all of the various parts worked together.

This approach look good to the outsider, but in practice it led to some faulty thinking. Although the systems engineers were selected from the best and brightest, an underlying premise of systems engineering was that the systems engineers didn't really need to be concerned with, and in fact shouldn't be concerned with, the details of the actual *design* of the system that they were managing. They just had to be concerned with the *process*. The fact is, a really good systems engineer—a top-notch systems designer—does have an encyclopedic knowledge of subsystem details, based on personal experience, and has enough confidence in his own abilities and knowledge to trust the people working on and solving the day-to-day engineering problems. He, in turn, earns the trust of the engineers under him by working closely with them, listening to them, and helping them to arrive at a successful design. But engineers are merely human, they are fallible. And this was the problem that systems engineering was supposed to address: if the project was being supervised by fallible men, then the generals, admirals, congressmen, and staffers who were overseeing the project, and who were responsible for the defense of the country, would be justified in doubting, questioning, and second guessing every decision made by the engineering managers. Systems engineering would eliminate the human factor.

The first step in systems engineering was to lay out requirements, starting with the top-level systems requirements. From those they would derive an ever-expanding list of more and more detailed requirements, eventually reaching into the smallest aspect of the design. These requirements would control the design process all the way down to the lowliest engineer. And, because the design was generated by one all-encompassing, integrated, state-of-the-art, modern—and purely analytical—process, there was no way to second guess the results at any step. Everything was seamlessly integrated, everything came out the only way it could. Everything started at the top and flowed with the requirements. When you used systems engineering, there just wasn't room for lapses in judgment, wrong decisions, unnecessary blind alleys, or failures. There was, therefore, no need to interfere in the management of the project.

Specific tools were invented to put a veneer of scientific rigor on the inevitable judgment calls that had to be made in designing a new system. Program planners wanted it understood that their design was not driven merely by experience, investigation, and experimentation. If the design was based on human judgment, it could be doubted. But, if it was based on science, it was beyond questioning. With the aid of these tools, the design decisions could be justified as the best that ever could have been made because you had scientific evidence to prove it.

One of the most powerful of these tools was the trade study. It worked like this: the systems engineers, working with the design engineers, came up with an approach that, based on their experience and investigations, they thought was most likely to work. They then engaged in an imaginative game of listing every possible design approach they could think of, no matter how far-fetched. The preferred approach and all of the alternatives were then analyzed according to a complicated equation involving many parametric constants and weighting factors for the requirements. This analysis generated so-called figures of merit for each option. The constants and the equation were then tinkered with until the chosen design approach came out as clearly superior.

When the design concept was presented to the top brass or congressional committees, it was presented as the best solution out of all possible solutions, arrived at by a purely scientific process. This made it essentially impossible for the design decision to be questioned. If there was any questioning of the design, the program managers would launch into interminable explanations and arguments concerning all of the many different design concepts, the figures of merit, and the trade table, until the questioner was exhausted. Thus any attempt to interfere with the decision that the engineers had already made could be blocked.

Other similar techniques came to be employed. The work breakdown structure, which was based on the idea that a few key decisions made at the top level would flow down to control all of the engineers at the bottom, was used to show how the systems engineers were in complete control of all decisions made. This hid the somewhat messy real-world process, whereby designs are developed by an upward flow of creative ideas and approaches that form the grist for the engineering judgment of the chief engineer. Of course, some of these tools did actually turn out to be helpful for thinking through complex programs. But they were no substitute for competence, judgment, and trust.

The PERT technique for scheduling was intended to show that, through the application of modern scientific management principles, yet another bugaboo of complex programs—the master schedule—had been tamed. It was also used to eliminate any taint of human judgment or management responsibility. Setting and following a schedule for a program that was blazing new trails in developing systems the likes of which had never before existed inevitably required a lot of engineering judgment, not to mention the application of copious amounts of money just to keep the project moving. But with PERT, blame or fault finding for the unavoidable schedule slips and resulting cost increases could be deflected from the managers.

Finally, the traditional process through which a good engineering manager established and maintained mutual trust and effective communication with the people working for him was shielded from scrutiny by a series of elaborate weekly status charts. These were used to show that management knew about every detail of the program and was carefully tracking its progress, independent of any reliance on trust or personal relationships between managers and engineers.

The whole charade was actually very effective and was even hailed as a major contributor to the success of the missile program. Nevertheless, when the truth is tampered with, bad consequences will always follow. Over time, the engineers and managers began to believe in the myths they had fabricated. As long as their systems were merely window dressing to deflect outside oversight, the real engineering could go on unimpeded. And although it was all recorded in the history of the program, it was naturally not spoken of widely. When Tom joined the company in the early 1980s as a young design engineer, the concepts of systems engineering and modern program management had seemed wrong-headed. With a little research he was able to uncover the myths on which they were based.

Nevertheless, when the truth is tampered with, bad consequences will always follow.

At that time, the senior program managers, if asked, would admit that they didn't let trade studies dictate design choice. Rather, they used the trade study as it was intended—to justify and defend their decisions. Unfortunately, the younger engineers, especially those labeled systems engineers, were enamored with the idea that by acquiring this title, they were suddenly experts in the design of systems. The title systems engineer had originally been given only to the most experienced and knowledgeable designers, who could function effectively in an oversight role, making sure the disparate parts of a project functioned together. This new crop of systems engineers were created by fiat.

Some engineers were assigned to the design groups, some were assigned to the systems groups, and some were assigned to the program office, whose primary function was handholding the customer. All three groups fought among themselves as to who was in charge, which tended naturally to get in the way of good engineering. But the most poisonous legacy of the whole systems engineering philosophy was the idea that someone who didn't really understand the details should nevertheless be in charge of making the big decisions in developing a design. And the younger systems engineers really believed good design was achieved by trading a multitude of conceptual designs.

Each component of systems engineering had over the years developed and taken on a life of its own. Carefully selected tasks were put forward to be tracked in the weekly status reports, while real issues were hidden, or worked out quietly on the side. Scheduling—which had always been a key management function in any large project—was handed over to annoying busybodies who built palaces in the clouds, insisting that everyone predict with unknowable precision exactly what needed to be done, when it needed to be done, and how long it would take. The schedulers, desperate for all of the inputs their computer programs could handle, would bring around 6-inch stacks of schedules for the engineers to check and revise on an almost daily basis. At one point, Tom's manager, in order to ensure that *some* real work got done, had limited scheduling to no more than two hours a day!

The antidotes for the mythology, waste, and obvious failures of "scientific" program management are trust and delegation of authority. Forsyth lacked Tom's direct experience in aerospace, but he knew that good people left alone, with sufficient resources, and headed in the right direction were going to do as good a job as could be done. His job was to supply the money, keep the investors off their backs, and be sure there was a market for their creation when they got done. For the vehicle itself, Forsyth was going to have to bet on one man, not a dispassionate process; he trusted Tom, and was willing to give him the authority to get the job done. Tom, for his part, had the trust of his engineers, trusted them in turn, and was willing to give them the authority to do their jobs.

With this in mind, Forsyth asked Tom to lay out his best shot, and he was either going to go with it to his last dollar, or fire him and get somebody else. Of course, Forsyth had some inkling as to what Tom thought was the best design approach—not all of Tom's exercises had been on way-out concepts. Some of them had, in part, been intended to build a consensus among his engineers about the direction the design would have to take. They had explored such issues as how much weight wings would add to a stage, whether it was possible to fly the first stage of a winged vehicle back to the launch site without excessive penalties, how best to design the trajectory for a two-stage vehicle, etc. The design team had already done a couple of iterations on Tom's favored concept; Forsyth gave Tom one month to put it all together and make a presentation, first to himself and then to the Group of Seven. When the time came, Forsyth had approved of the concept, and took Tom to the Group for the big pitch.

CHAPTER 7
The Pitch

FORSYTH brought Tom in alone to give the presentation. The Group of Seven agreed to hold questions until the end, and six of them settled back to listen. The seventh—Forsyth—was on the edge of his chair as Tom began.

"For a conventional, vertically launched ballistic vehicle—with no lift-producing wings—the path the vehicle takes to get to Earth orbit must start out with the vehicle thrusting with a thrust-to-weight ratio greater than one, or it doesn't get off the pad. Typical designs will find an optimum value somewhere in the range of 1.15 to 1.2. After liftoff, the vehicle will climb vertically for 10 to 20 seconds, at which point it will turn slightly to the east, at just a few degrees from vertical.

"The vehicle then follows what is known as a 'gravity turn' trajectory, along which the vehicle flies at zero angle of attack. This means that it flies in the direction it is already going, so that the relative wind is nose-on. However, the force on the vehicle due to gravity, acting toward the center of the Earth, has a component perpendicular to the direction of flight, because the vehicle is not going straight up. This force component slowly turns the vehicle.

"In other words, if you look at the trajectory second-by-second, at the end of each second, gravity has slowed the vehicle down by an amount equal to the acceleration of gravity—*g*—times the sine of the flight-path angle (measured from the horizontal) times one second. Gravity also adds a velocity component equal to *g* times the cosine of the flight-path angle times one second. Of course, the vehicle is also being accelerated by the rocket engine thrust, acting in the direction of the flight path during the interval. When both of the velocity changes due to gravity and to the engine thrust are added vectorially to the velocity at the beginning of the one second interval, the net change in the magnitude of the velocity vector (the 'speed') is the thrust divided by the vehicle mass minus the sine of the flight-path angle times *g*, all times one second. The 'cosine of the flight-path angle times *g* times one second' component, since it acts in a direction perpendicular to the flight path, causes a change only of the direction of the velocity vector, not its magnitude, thus decreasing the flight-path angle.

"The length of the vertical portion of the trajectory and the timing of the 'kick' over to the gravity turn are chosen so that, in theory, at burnout of the final stage, the vehicle will have a zero flight-path angle (i.e., its trajectory is parallel to the surface of the Earth), and an altitude well above the atmosphere—somewhere between 60 and 100 nautical miles. There is no simple equation that will spit out the most efficient path to take to orbit, nor is it possible even to simply calculate how much thrust should be used or how long it should be applied. The integration of the thrust along the flight path is highly dependent on the path itself. If the rocket thrust is too high, the vehicle will reach orbital speed too low in the Earth's atmosphere, resulting in excessive heating and drag. Or, it will burn out with a flight-path angle that is greater than zero, meaning that the vehicle enters a highly elliptical orbit (which typically intercepts the Earth). It's not even easy to determine exactly how much velocity the vehicle needs to get to orbit, either.

"The rocket equation delta-V $= Isp \times 32.2 \times \ln(MR)$ characterizes the performance of a vehicle in free space, far away from any gravitational forces, and without atmospheric drag. It tells you, for those conditions, what change in velocity a vehicle would achieve, for a given mass ratio *MR* (the ratio of the weight of the vehicle at launch to its weight at burnout) and a given (constant) exhaust jet velocity ($Isp \times 32.2$ ft/s). But you cannot analytically calculate the total velocity change (assuming you start from zero at the Earth's surface) required to achieve orbit. What you can calculate is the orbital velocity for a given orbital altitude; for example, at 100 nautical miles, orbital velocity is about 25,500 feet per second.

"The next step is to find out how much additional velocity will be required to overcome the real-world atmospheric drag losses, gravity losses, steering losses, and other losses that will have to be made up if your final velocity is going to be the desired orbital velocity. These losses must be added to the orbital velocity to determine the total velocity change that your vehicle will have to produce. The usual method for determining these losses is to simply follow the vehicle and see what happens, using a special-purpose computer application called a trajectory program. Step by step, such a program calculates all of the forces acting on the vehicle over specific time intervals all the way from liftoff to orbit insertion, using appropriate vehicle characteristics (measured or estimated) such as drag coefficients, aerodynamic coefficients, engine operating parameters, atmospheric models, etc. The net force is then divided by the mass of the vehicle (which is, of course, always changing as propellants are consumed) to get the net acceleration, which is then integrated to get the velocity and the position of the vehicle at the end of each time interval. The various trajectory parameters—vertical rise time, kick angle, engine thrust profile, and others—are varied in the program until you get a result that says that the vehicle reaches the proper altitude and flight-path angle with the proper velocity.

"At the end of this iterative process, you will find that, instead of a velocity change equal to the orbital velocity of 25,500 feet per second, a total equivalent velocity change (in free space) of somewhere between 30,000 and 31,000 feet per second is required, depending on specific vehicle design characteristics.

"ALMOST everybody who has ever looked at launch to Earth orbit agrees that a single-stage vehicle would be desirable. Almost all other vehicles, be they car, airplane, or ship, consist of a single stage. If you want to go farther than your car can go, you stop and refuel it. Transportation is broken up into segments, or legs, each of which are within the capability of the vehicle being used. The problem with space launch is that you can't refuel halfway to orbit. You have to go all the way or nowhere at all. And, unfortunately, the velocity requirement of 30,000 feet per second or more makes it virtually impossible—with existing technology—to build a single stage that can place a payload of more than about 3% of its gross lift-off weight (GLOW) into orbit.

"It also happens that 3% of the GLOW (or about 30% of the dry, or empty, weight of the stage) is about what you need for the subsystems that allow the vehicle to return to Earth in a condition to be reused again. These include retro rockets to slow the vehicle out of its orbit for reentry into the Earth's atmosphere, a heat shield, and a soft landing system. In addition, building an airframe and propellant tanks suitable for many hundreds of flights is inevitably going to produce a vehicle somewhat heavier than any existing single-use expendable rocket.

"Many would-be launch vehicle designers looked on single stage as the only way to go. This led to a back-and-forth debate over whether or not it is essential for low costs and, if it is essential, whether it can be accomplished. Most of the aerospace establishment concluded in the 1990s that it was essential, followed by a conclusion that it wasn't possible with existing technology. Then came a continuing search for exotic new technologies that would make it possible. After all, all you have to posit is a 30% drop in the weight of all the components of the vehicle, or a 20–25% increase in *Isp* for the engines.

"Unfortunately, 40 years of experience have not shown any significant downward trend in the weight of most of the basic components of a launch vehicle. The weight of metallic propellant tanks has proven to be about as low as it can go; it's basically a function of the propellant density. Composite tanks are often touted as a solution, but the difficulties of fabricating large composite tanks are not easily dismissed, and by the time all of the attach points, plumbing fittings, access ports, and other details required in a working tank are added, the weight of the composite structure ends up pretty close to the weight for conventional metal construction. Most other component weights are unlikely to yield much improvement, either. Yes, electronics continue to get smaller and lighter, but

"Unfortunately, 40 years of experience have not shown any significant downward trend in the weight of most of the basic components of a launch vehicle."

that will only go so far—those components do not represent a big chunk of the total weight.

"In terms of *Isp*, well, the theoretical performance limits of rocket engines—using practical propellant combinations—have already been fairly closely approached. Therefore, much effort has been devoted to finding a suitable alternative to rockets. Airbreathing engines have received the most attention. They produce a higher thrust for a given amount of propellant (higher specific impulse) because they only need to carry the fuel; the oxidizer is the oxygen in the surrounding air. Variations have been proposed, such as one that collects the oxidizer for a rocket engine by liquefying the air through which it flies. However, airbreathing systems have fundamental problems. Even though they have much higher specific impulse than rockets, a typical airbreathing engine has a thrust-to-weight ratio of only seven or eight, maybe 10 for some purpose-built examples, so that the weight of the engines alone would be nearly equal to the weight of the entire structure of a rocket-powered single-stage vehicle. The other major problem with airbreathing engines is that, not surprisingly, you have to stay where the air is. That means that your vehicle will experience increased drag, which goes up as the square of the velocity, and increased aerodynamic heating, which increases as the cube of the velocity.

"Since so much debate has been generated about airbreathers in the past decade or so, it is worth digressing for a few minutes to talk about the only high-speed aircraft that has ever flown operationally, the SR-71 Blackbird. This remarkable vehicle provides some useful insight into the difficulty of flying an airbreathing aircraft at high Mach numbers. The SR-71 had its genesis in a military program to build a hydrogen-powered jet airplane, and the Air Force turned to Kelly Johnson and the Lockheed Skunk Works for the design. It didn't take Johnson and his team long to determine that the liquid hydrogen fuel, with its extremely low density, would have a volume of over five times that of an equivalent amount of higher density kerosene jet fuel. (The density of liquid hydrogen is just under 4.5 pounds per cubic foot; high-density kerosene jet fuels such as JP-7 have densities over 60 pounds per cubic foot. Liquid hydrogen does have three times the energy per pound as kerosene, so that the required volume is five times greater for the hydrogen.) However, when burned in a jet engine, hydrogen produces only 20% more thrust. So you end up with an airplane which has one-third the fuel weight and 20% percent more thrust, but it requires a fuel tank five times larger. This means that as velocity increases, the much bigger hydrogen-powered airplane soon loses to drag any advantage it gains in thrust or lighter weight. When he had figured this out, Johnson gave the money back to the Air Force and cancelled the program. Eventually, this program led to the JP-7-fueled SR-71.

"The SR-71 was originally designed to fly at Mach 4, or four times the speed of sound, about 3000 mph. But, as far as anyone could tell, it was actually only capable of Mach 3.2. This is a good indication of the difficulty of overcoming the drag and heating problems as you try to make an airplane go faster and faster. It has been over 40 years since the SR-71 first flew, and although no one can say for sure whether or not the Air Force is flying a faster, classified vehicle, it seems unlikely that anybody has ever gotten anything to fly at sustained speeds greater than Mach 4. Yes, NASA flew the X-43A, which maintained a speed of Mach 10 for

10 seconds burning hydrogen fuel, and a missile and an artillery projectile are under development using scramjet technology, but there is no one in the industry who believes in a scramjet powered launch vehicle. Some $3 billion were spent on the National Aerospace Plane (NASP) in the 1980s and 1990s, with endless experiments in supersonic wind tunnels and shock tunnels, but nothing that approached a practical vehicle.

"No, there isn't much evidence to support the idea that an airplane can be built to fly at Mach 6 or 10, much less at the Mach 25 speeds necessary for orbital flight. Combining airbreathing engines with rocket engines tends to make things go from bad to worse. To optimize the rocket part, you want to exit the atmosphere as quickly as possible, which doesn't give the airbreathing part much opportunity to work. And, as we have already discussed, the airbreathers tend to be much heavier for the amount of additional thrust provided. The additional weight of the wings that this hypothetical spaceplane would need of course exacerbates the problem of building a vehicle that is light enough to reach orbit with a single stage.

"Combining airbreathing engines with rocket engines tends to make things go from bad to worse."

"IF a single-stage vehicle is not technically feasible, at least within our budget and time constraints, madam and gentlemen, and a hypersonic airbreather is even farther out, then attention naturally turns to a two-stage vehicle.

"Adding stages to a rocket has the effect of increasing the mass ratio, which if you will recall the rocket equation once again, means that the delta-V capability is increased. I like to think about staging as a process of continuously decreasing the weight of the structure, engines, and other nonpayload items as the vehicle flies; as propellants are consumed, of course, the tanks don't have to be as large, and less thrust is needed from the engines. If it were possible to build a vehicle in which the weight of the tanks and engines amounted to 10% of the gross weight, not counting the payload, and from which structure could be discarded in infinitesimally small increments, this continuous—or infinite—staging would have the same effect as diluting the vehicle propellants. Thus, even though staging can be used to increase the mass ratio, the maximum performance gain that can be achieved is limited by the effective decrease in *Isp* by a percentage equal to the structural mass fraction of the vehicle.

"So, a vehicle with a 10% structural mass fraction, a nominal *Isp* of 300 seconds, and infinite staging can be considered equivalent to a single-stage vehicle with an *Isp* of 270 seconds, and no structure weight other than the payload; the amount of propellant needed to place a pound of payload and vehicle into orbit would then be equal to (*MR*-1), where *MR* is the mass ratio. As you use more and more stages, the performance of the vehicle approaches the infinite staging case. In the early days of rocketry, when staging was not well understood, many engineers and managers were startled by the increase in payload that could be gained by adding another stage to an existing launch vehicle. Staging, like the launch trajectory—to which we will return shortly—is not easily optimized by purely analytical methods, because of the incremental cost and weight associated with adding another stage. There are also some aspects of staging which are not immediately intuitive. For instance, if one imagines a vehicle with a structural weight of only 4% of gross vehicle weight, and a payload of 8% of gross vehicle weight, infinite staging would only increase payload to 11%, or by less than a third. Practical staging would only increase payload to about 10%. And this one may be particularly surprising—adding an additional stage to a well-optimized three-stage vehicle will provide little more benefit than adding the equivalent weight of propellant to the first stage. You can't always squeeze a little more performance out of a vehicle by adding one more stage.

"But getting back to the issue of trajectory design, two stages presents a whole new set of problems. First of all, because the first stage provides only part of the velocity necessary to achieve orbit, it must inevitably fall to Earth some distance away from the launch site, typically 50 to several hundred miles. And the first stage won't always come down in the same spot, if you are launching into different orbital inclinations on different flights.

"The fact that the stage does return to Earth down range also means that flight over populated areas is problematic. Launching out over the ocean has been the solution for American rockets, but that restricts launch sites to the coast. Also, if you are going to try to recover your first stage for reuse, water landings will contaminate the vehicle with salt water, which inevitably complicates the design, shortens the operational life, and compromises reliability, due to corrosion effects. There is also the possibility of losing the stage to storms at sea, unless you restrict launches to good weather. And a ship would be required to return the stage to the launch site. Conceivably the ship could be prepositioned down range so that you could land the stage on it. However, this approach requires not only good targeting or piloting, but also the added expense of a special purpose ship. Either method requires waiting for the stage to be brought back. For these reasons, a basic requirement of our design is that both stages return directly to the launch site, under the control of a human pilot. This will enable our customers to operate their vehicles from anywhere in the world, and eliminate any concern about the stage falling in populated areas.

"For these reasons, a basic requirement of our design is that both stages return directly to the launch site, under the control of a human pilot."

"So, how do we get the first stage back to the launch site? The most obvious method would be to put wings on it and fly it back. However, it is difficult to make a first stage flyable. The heaviest part of any stage tends to be the engines, which are almost by necessity located in the rear. A flying vehicle needs to have its center of gravity well forward, if you want it to be stable. The wings also add considerable weight, as well as complexity, because even in the most basic concept, you are dealing with a hypersonic glider. Hypersonic vehicles require a lot of analysis and testing, and hypersonic wind tunnel testing is expensive. The shuttle required 99,000 hours of wind-tunnel testing, and when the Europeans undertook the development of their Hermes vehicle, they figured it would tie up every major tunnel in Europe for years. Furthermore, no one has ever built a flyback first stage, so we'd be treading new ground.

"Another approach, used in the Kistler K-1, is to carry sufficient propellant onboard the first stage so that after second-stage separation, the first stage—now considerably lighter—can be flown on a ballistic trajectory which targets the launch site. This is actually a variation on the solution which I am going to set forth shortly. But it does require considerably more propellant weight and volume in the stage, and it relies on restarting one or more engines. It might be possible to avoid restarting the engines by gimbaling them until they are perpendicular to the flight path, so that the thrust is momentarily nulled out during second stage separation, and then gimbaling them back for the retro maneuver. There are also other issues for two-stage vehicles. Staging within the atmosphere presents problems, and escape systems for the second stage are more difficult, but these are problems to which engineering solutions exist.

"LET me return now to the powered flight trajectory, so that I may explain the staging approach which I am going to propose. I'll start with a question. If the first stage is launched completely vertically, how much velocity can it usefully add to the second stage? Remember, typical trajectories have brief vertical segments, followed by a gravity turn shortly after liftoff. Obviously, then, there is a limitation on how much velocity the first stage can constructively add if all of its velocity is in a vertical direction. Boeing, during the Heavy Lift Launcher studies of the early 1970s, looked at various options to deal with the problems of first stage recovery. One of these was a completely vertical trajectory for the first stage so as to allow it to be recovered at the launch site, instead of downrange. However, Boeing was considering a two-stage vehicle design that had been optimized for minimum launch weight, and calculated a 70–80% reduction in payload for flying the first stage completely vertically. Clearly, launching the first stage vertically requires a design configured to the trajectory.

"Let's turn now and look at the velocity, the delta-V, that the upper stage would need to produce in order to achieve orbit, if the first stage provided all of the vertical velocity that could possibly benefit the upper stage. Right away, we can see that the second stage must provide all of the horizontal velocity needed. As I mentioned earlier, orbital velocity at 100 nautical miles is 25,500 feet per second, but here we can get some immediate relief if we consider an eastward launch. Because the Earth rotates once about its axis every 24 hours, and the circumference of the earth is approximately 24,000 miles, the Earth's surface is moving eastward at 1000 mph, at the equator. So, depending on the latitude of the launch site, an eastward launch decreases the velocity needed by 1200 to 1400 feet per second. Thus, the total velocity change required by the second stage is reduced to approximately 24,100 feet per second.

"If the rocket equation is now used with this lower velocity requirement, approximately 18% of the stage gross weight will reach orbit. If 5% of the second stage gross weight is payload, then 12–13% is available for the structure and engines. A structural mass fraction of 12–13% is right in the range where historical experience and the consensus of the experts would agree that a reusable stage can be built.

"Furthermore, in the proposed concept, the second-stage engines always operate in a vacuum, thus allowing relatively low-pressure engines to achieve high *Isp*. It also means that the second-stage will experience little or no aerodynamic drag losses. This simplifies the design, because building a small and relatively high-drag orbital stage will not penalize its performance. And, because there are no atmospheric loads during second-stage flight, the second stage will require less engine gimbaling and incur lower steering losses. Finally, the propellant tanks won't require slosh-baffles, thereby reducing the structure weight.

"Thus, madam and gentlemen, the vehicle that I propose to build is a two-stage vehicle, with a first stage that is launched vertically, remains essentially directly over the launch site during its flight, and then lands back at that site. We could reduce the total velocity required by the second stage even further by providing some horizontal component of velocity from the first stage. But that would mean that the first stage has to fly downrange, and in order to bring it back to the launch site we would have to use a more complicated vehicle and trajectory—something like the Kistler 'pop-back' maneuver. The downrange or horizontal component must thus be traded against increased system complexity and greater weight for the first stage. Given that a completely vertical trajectory will still allow us to design a second stage that we can actually build with available materials, technology, and techniques, the use of the more complicated first stage does not appear to be justified. However, we will have an option for increasing the payload capacity for special missions by flying the first stage down range.

"We're still not done talking about the powered flight trajectory. The one we're going to use with this bird is somewhat odd. As noted earlier, the total velocity required for an eastward launch is about 30,000 feet per second, and, as we've seen, about 24,100 of those feet per second directly contribute to the final, orbital velocity of the second stage. Where do the other 6000 feet per second go? Well, they go to make up the losses that we've talked about. There are four major contributors; in order of importance they are gravitational losses, aerodynamic drag losses, and steering losses. The fourth one, which is really not a loss at all, is the gain in potential energy that the vehicle gets as it goes from the Earth's surface to an altitude of 100 miles above it. The velocity loss attributed to the potential energy gain can be seen to be the velocity needed to transfer from an orbit at the Earth's surface to an orbit 100 nautical miles higher; it turns out to be about 370 feet per second. The aerodynamic drag losses depend on the vehicle's drag characteristics. Drag, of course, is the force of the atmosphere—actually the relative wind caused by the motion of the vehicle through the atmosphere—pressing against the vehicle. The Saturn V, for example, experienced drag losses of approximately 185 feet per second, but for a small vehicle like the one we're talking about, they will be in the neighborhood of 600–1000 feet per second.

"Steering losses result from the need to use engine thrust to keep the vehicle under control and on course, which means that the engine thrust vector is not always perfectly aligned with the velocity vector of the vehicle. This misalignment of the thrust vector decreases the velocity that is added along the flight path, resulting in a loss in performance.

"The most significant losses are those due to gravity. As I mentioned earlier, gravity losses are approximated by multiplying the sine of the flight-path angle times the force of gravity times the burn time of the engines. In other words, the acceleration, or change in velocity that you get from your engine thrust is reduced by an amount equal to the product of those three quantities. Of course, even if the engines are not operating, gravity will continue to affect the vehicle. But here I'm concerned with the delta-V that we actually get from the thrust of our engines during the time that they operate, so we have to subtract the effect of the gravity force during that burn time. Now, even though the force of gravity decreases the farther you get from the center of the Earth, the vertical trajectory of this vehicle can result in relatively high gravity losses.

"One standard method for minimizing these losses is to increase the thrust-to-weight ratio—the ratio of the engine thrust at liftoff to the vehicle weight at liftoff—so that the engines get their job done faster, with less time for the gravity losses to have their effect. For this reason, I plan to use a T/W of about 1.6 for the first stage, which is rather high compared to most conventional launch vehicles. Thus, at launch, 60% of the thrust supplied by the engines will be available to accelerate the vehicle. As the engines consume propellants and reduce the weight of the vehicle, the thrust-to-weight will increase, reaching a maximum g force of about 4 just before the first-stage main engines shut down. The first-stage engines will be of a new design, but based on a yet-to-be-chosen existing engine—possibly the Boeing Delta first-stage engine, or even the old H-1 from the Saturn 1B. The engine we select, reengineered for reusability and running on liquid oxygen and liquid methane—natural gas—will have relatively high performance. The propellant costs will be in the neighborhood of $0.07 per pound.

"Unfortunately, this high thrust-to-weight presents another problem. It means that the first stage will burn out at a rather low altitude of approximately 100,000 feet. This is just above the altitude of the highest flying airplanes, and with the vehicle traveling at about 2700 feet per second, it will experience significant drag forces. High drag forces complicate staging, and staging at this altitude would also require the second stage to fly while being subjected to considerable aerodynamic loads. To eliminate those issues, four smaller, sustainer engines will be used to burn the last 5% of the first-stage propellants, at a thrust-to-weight of about 0.3. At first it would seem that thrusting vertically with a thrust to weight of less than one would accomplish nothing because the force of gravity would be greater than your thrust. However, such a conclusion is based on an erroneous conception of gravity losses. You must also take into account the work done by the thrust times the distance over which the thrust is acting. So, even though this thrust will increase the velocity of the stage by less than the gravity force decreases it, because of the vertical velocity, considerable work will be done on the stage.

"Let me use the following thought experiment to give you an intuitive understanding of what this means. If we take a rocket that is positioned 1 foot over the launchpad and burn all of the propellant at a thrust to weight of exactly one, the vehicle will go nowhere—all of the possible delta-V will have been matched or cancelled by the gravity losses. Now, take another vehicle 1 foot above the pad, also with a thrust to weight of one, except now we give it a vertical velocity of 1 foot per second. If this vehicle were capable of thrusting long enough, it would eventually be so distant from Earth that its 1 foot per second velocity would be equal to escape velocity at that point in space. Now we're getting somewhere! So, even though there is no net thrust to accelerate the vehicle—nothing, in other words, to give it some delta-V—the work done on the vehicle by the engines will give it a big increase in potential energy.

"The purely vertical trajectory of the first stage means that the second stage, after it separates from the first, has only vertical velocity—with no horizontal component. If that vertical velocity is large enough—and it only needs to be a rather modest 3000 feet per second—the second stage can be then be flown with a zero flight-path angle, in a purely horizontal direction. That means that the gravity force is acting perpendicular to the flight path, so that the acceleration due to gravity will, as the second-stage engines burns for some 260 seconds, 'eat away' at only the vertical, upward velocity provided by the first stage. The first stage will of course also have an upward velocity of about 3000 feet per second at staging; this will cause the first stage to continue coasting upwards for about 100 seconds, before it begins to fall back to Earth.

"It would thus appear that the second stage would require a high enough thrust-to-weight ratio to allow it to get to orbit within that 100 seconds. However, as a vehicle approaches orbital velocity, the force of gravity is counterbalanced by the centrifugal force, and so affects it less. If you will recall, when an object is in Earth orbit, it is constantly being accelerated towards the Earth due to gravity, but its horizontal velocity carries it forward along the curvature of the Earth. The net effect of gravity then is to simply 'bend' the otherwise straight-line trajectory around the Earth. An orbiting satellite—in so-called 'free fall'—is not uninfluenced by gravity; the combination of its forward velocity and the acceleration due to gravity together cause it to move in a circular path around the Earth.

"So, actually, if we use a thrust-to-weight of one, the vehicle will not reach orbit by the time it begins to fall back into the atmosphere, unless we fly with a slight positive flight-path angle. It turns out that a thrust-to-weight of about 1.4 will be needed to achieve orbit for a second stage with a delta-V capability of 24,500 feet per second and a payload of 5000 pounds. However, the trajectory is not highly sensitive to thrust to weight; a value of 1.0 would result in only a 20% decrease in payload, down to about 4000 pounds.

"But, as mentioned earlier, even though the first-stage main engines burn out at approximately 100,000 ft altitude, the last 5% of the fuel will be burned in four lower-thrust sustainer engines, for a period of about 44 seconds. This low-thrust coast period will, even at a thrust-to-weight of only 0.30, reduce gravitational losses during the burn time and do some useful work on the vehicle as it climbs to 200,000 feet before staging. At that altitude, the atmosphere is thin enough that any aerodynamic forces on the vehicle will be insignificant. The lower thrust level also allows us to maintain that thrust for positive control of the vehicle until staging has been accomplished.

"SO, gentlemen, madam, we've chosen a two-stage vehicle, but one in which the second stage is very much like a single-stage vehicle. And with the first-stage configuration that we have selected, we have reduced the velocity requirement for the second stage down to a point where the required mass ratio is well within the capability of available technology. I don't think that anyone in the industry could say that such a stage cannot be built.

"That said, we will still be using every trick we can to achieve the weight targets with an elegant design. The upper stage will have a mass ratio of 5.8, with a GLOW of 99,000 pounds and a weight at burnout of 17,000 pounds. The first stage mass ratio will need to be only about 2, depending on the exact performance of the engines; it might go as high as 2.5. However, because of the low staging ratio, over 30 pounds of structural weight could be added to the first stage with the loss of only 1 pound of payload to orbit. Thus, the first stage is highly insensitive to weight growth. Our initial design point is for a structural mass fraction of about 19%, which means that the stage structure and subsystem weights can be three to four times higher than those of any well-designed expendable. That gives us a lot of margin.

"The weight insensitivity of the first stage simplifies its recovery options. As the DC-X demonstrated, vertical landing under rocket power is a relatively straightforward proposition. The sustainer, or vernier, engines are of such size that they could be restarted as the vehicle approaches the ground, for a powered landing. Alternatively, vertically thrusting jet engines could be used. We intend to investigate both approaches. Both techniques have been proven, and are well within the state of the art.

"The first stage will return to Earth from about 350,000 feet altitude, and will reenter the atmosphere with the same velocity at which it left it, or about 3000 feet per second. This results in very little aerodynamic heating, not much worse than SpaceShipOne. The thermal protection system, therefore, need consist only of making sure that the parts of the vehicle that will be exposed to direct heating have sufficient mass so that in the relatively short time of exposure they won't heat up significantly. And—because weight is not very critical on the first stage—the quick solution for any heating problems that may crop up can be as simple as increasing the mass of the structure to absorb more heat, or spraying water on or running water through any parts that tend to get too hot.

"Now, on to the second stage. As you can see in this rendering, the basic shape of the vehicle is conical. The oxygen tank is at the forward end, meaning that most of the mass is forward—which makes for a more stable vehicle—and the larger hydrogen tank is at the rear; its large-radius lower surface forms the structure backing the reentry heat shield. Submerged in the hydrogen tank is the propulsion module, consisting of six off-the-shelf Pratt & Whitney RL10 engines. These are actually the RL10A-4-2 model, with a vacuum thrust of 22,300 pounds. They cost about $3.5 million each, or $21 million a ship set. Pratt & Whitney also has another flight-qualified model—the RL10B-2—with approximately 24,000 lb of thrust, although it's not currently in production. With the RL10A-4-2, or RL10B-2 our vehicle will place over 5000 pounds into orbit; with the lower-performance production engines, RL10A-3-3A we'll get about 4000 pounds.

"So, the vehicle consists essentially of two propellant tanks, separated by an interstage section. In this section, we will place our payload bay and pilot station. By doing this, the additional weight for a payload bay is reduced to a few hundred pounds. The only extra structure needed is just what it takes to make the interstage between the tanks a little longer than would be necessary if you didn't put any payload in there. The shape of the payload bay is not conventional, nor particularly convenient. It is in some ways spacious, though, with 1100 cubic feet for 5000 pounds of payload. However, some of that volume will be taken up by the pilot's ejection seat, the altitude control system, and the guidance computers.

"The entire payload bay is pressurized, so deploying a payload requires venting the bay. The pilot will manually deploy the payload through a hatch—wearing a spacesuit, of course. Because our goal is to provide low-cost spaceflight with a manned vehicle, we are willing to compromise the convenience of the payload. Obviously, redesign of many current, typical payloads—including redesign for on-orbit assembly for the more complex and larger payloads—will be required

"Because our goal is to provide low-cost spaceflight with a manned vehicle, we are willing to compromise the convenience of the payload."

for them to fly on our vehicle. However, in view of the costs we expect to achieve—around $200 per pound—this vehicle is going to have to be much more like an airplane when it comes to loading cargo. If you want low cost, you'd better be willing to put your payload in a standard shipping container.

"The vehicle will reenter the Earth's atmosphere tail first—ballistically, in other words—like a projectile, not like an airplane. Actually, a ballistic vehicle can have a small lift/drag ratio, somewhere around 0.1, which can moderate the reentry *g* loads to five instead of eight *g*, and also provide a small amount of cross-range capability to help compensate for uncertainties in atmospheric conditions during reentry. This technique has been used on the Apollo command module as well as the Russian capsules.

"WE are looking at several concepts for the heat shield. The large-radius spherical shape of the base will restrict serious heating to that area and—if we use a noncatalyzing surface—maximum temperature will be about 2600°F. That is well within the limits of the shuttle tile materials, but more importantly, it's within the limits of some very attractive advanced blanket materials developed by NASA Ames Research Center. The buried engine module or bay will employ either direct cooling of the engines by means of a hydrogen bleed through their cooling channels, or with a water spray. Or, we might go with a deployable heat-shield segment to cover the engine openings. If those approaches don't turn out to be workable in the near term, we could use water to cool the entire heat shield, boiling it off as it absorbs the heat. However, that approach would impose a weight penalty of approximately a 1000 pounds over our baseline weight of about 600 hundred pounds for the base heat shield.

"I'm going over a number of approaches, to show that in some of the trickier aspects of the vehicle, which have not been completely worked out, we do have several alternatives—some more or less risky. There are also some 'brute force' techniques that we could fall back on if we have to, while developing more elegant, lighter solutions.

"Probably the most controversial design decision in the entire vehicle is selection of the recovery mode for this stage. But before I cover that, let me talk for a few moments about design philosophy. My basic approach to our objective of low-cost space transportation includes a pilot who can land the vehicle on solid ground, like an airplane. I aim to focus on this feature using what I refer to as the 'Kelly Johnson approach' of accepting certain operational limitations that might at first seem unacceptable. An example of Kelly Johnson's approach is the U-2, which was built without any real advances in technology. It was simply a high-altitude glider powered by a jet engine, and it was highly successful in achieving the capability for high-altitude, long-range flight that was necessary for reconnaissance over the Soviet Union. However, in order to reach that capability, it ended up with what most designers would consider to be marginal landing gear, and it was extremely unstable in flight, requiring the constant attention of a very skilled pilot to keep it in the air. These limitations kept the U-2 from being considered a typical operational military airplane, yet it provided unusual capability for its time by accepting some compromises in operability to get the required performance.

"Even more so the SR-71. It was designed without any movable wing surfaces and had such a high minimum takeoff velocity that it could not take off fully loaded with fuel. It took off with just enough fuel to get it to a rendezvous for refueling from an aerial tanker. Again, not operationally very convenient, but it substantially simplified the design to the point where it could be built.

"The operational compromise for our vehicle is the use of a parawing for recovery and landing. Yes, that will require packing and repacking what is essentially a large parachute for each flight, and provides the possibility of damage due to abrasion from contact with the ground or other objects, and maybe even the hot sections of the vehicle after landing. But the lightest recovery system for a ground landing appears to be a parawing. This one will be the largest ever used on an operational vehicle. Fortunately, NASA's X-38 program of the late 1990s flight-tested an example capable of supporting 19,000 pounds. That is just over the empty weight of our vehicle with payload. A parawing can—with a landing flare—provide a vertical velocity of less than 15 feet per second on touchdown. That will allow the use of a relatively simple, lightweight landing gear, weighing only about 1.5% of the dry weight of the vehicle, for a safe, reliable, and reasonably comfortable landing. Our baseline actually includes two parawings, each with an estimated weight of 7% of the dry weight of the vehicle, on separate shroud systems. That way, if the first one fails to deploy properly, it can be cut away and the backup deployed.

"For recovery from catastrophic in-flight emergencies, the trajectory allows the use of ejection seats in both stages, with ejection conditions no worse than those of a high performance military aircraft. With relatively simple modifications to standard seats, we expect to be able to provide for recovery anywhere along the launch trajectory. Propellant dump capabilities will be incorporated on both stages, so they can be recovered through most of the flight envelope as well.

"With use of the off-the-shelf engines for the second stage, at a cost of $21 million per ship set, and estimating the total cost of the vehicle at three to five times the cost of the engines, the upper stage will have a production cost in the $60 to 100 million range. The lower stage, although heavier and requiring higher thrust engines, is expected to have a cost of about one-half that of the upper stage, or somewhere in the $30 to 50 million range. This puts the total estimated cost for the entire vehicle at $90 to 150 million, for the 20th production vehicle.

"In summary, our overall concept is a two-stage design that any expert in the industry would have to agree could be built and could achieve orbit, although they might quibble about what the exact payload will be. An all-metal structure is baselined. The vehicle does not use wings and can thus be designed with simple analytical extrapolations from existing flight test data for simple aerodynamic shapes. We will certainly want to do a bit of wind-tunnel testing, but we definitely won't need anything like the 99,000 hours of wind tunnel time that the shuttle used. The recovery systems are uncoupled from the main propulsion systems, which simplifies the design. Everything is recovered at the launch site, thereby simplifying logistics considerations.

"Many complex aerospace vehicles have proved to be capable of long, useful lives. Vehicles such as the B-52 bomber, the Boeing series of jet airliners, and even the space shuttle, have demonstrated operational service lives of 30 to 50 years. This vehicle has the potential of fulfilling a wide range of missions over many decades. For instance, with ejection seats for each passenger, it could be configured to carry five people into orbit at a time; without ejection seats, it could carry 10–20. It could also—with on-orbit refueling—transport 40,000 pounds to geosynchronous orbit and return the stage to low Earth orbit with the use of aerobraking.

"This vehicle has the potential of fulfilling a wide range of missions over many decades."

"Eventually the upper stage, with a delta-V capability of approximately 25,000 feet per second, could be refueled in low Earth orbit, and if modified for lunar landing, transport 5000 pounds to the lunar surface and return. It might generally be used throughout the inner solar system for transport to Earth-crossing asteroids and even—with refueling on the Martian surface—for transport between Earth and Mars. These possibilities need some further study, of course.

"Mr. Forsyth has asked me to also lay out for you a broad outline of how the project will be executed, should you agree to provide the financing. I propose a two-pronged program: the first to build a prototype vehicle using as much existing hardware as possible, with a two-year design and build phase and a six-month test program. The second prong will, in parallel, undertake the design of the production vehicle, and get it into production. The target for first flight is five years from go-ahead. This schedule is tight, but doable. I have been using some of that $50 million which you provided for operations over the past year to purchase long lead items for the production of RL10 engines, to make sure it is doable.

"My bottom-line goal for the prototype is to get it to orbit with a pilot on board and return safely to Earth, somewhere between five and 20 times. We will demonstrate reusability, flying one prototype as many times in a short period of time as possible, consistent with safety considerations. Not worrying about payload in the prototype will give us plenty of weight margin. The prototype will allow us to wring out the major design features, as well as the design teams, production teams, and flight operations and support teams and systems. It will also allow us to work through the regulatory hurdles before we are in production, and it should be able to convince potential customers that we are serious. At the same time, the flight test results will provide a level of comfort as we finalize the design of the production vehicles.

"I propose to build three sets of prototype flight hardware. We will probably fly the second set, the third set will never be completed, and the first set will be heavier than we want it to be. I think three sets are the minimum to ensure program viability, should we lose a vehicle during testing. Though our cost estimates are still rough at this time, we are looking at a budget of $1 billion to get through the first two and a half years—through flight test. This budget includes $100 million for engineering of the production vehicle, and we expect to come back for advanced funding of production facilities at about 24 months from go-ahead. Looking at a production rate of about eight vehicles per year, construction of the first production airframe will commence at the beginning of year four. Our current estimate for the production program cost—again, at an initial annual production rate of eight vehicles—is $2 billion, which will need to be committed about 24 months from go-ahead. Thank you."

The Group of Seven, as you would expect from a band of long-time space enthusiasts, had a lot of questions for Tom. Many of them had to do with the cost and business aspects of the program, but a number of the questions were technically very astute. Tom handled them all deftly. After the Group adjourned, Forsyth shot Tom a "thumbs up."

AM&M
AM&M

CHAPTER 8
The Business Plan

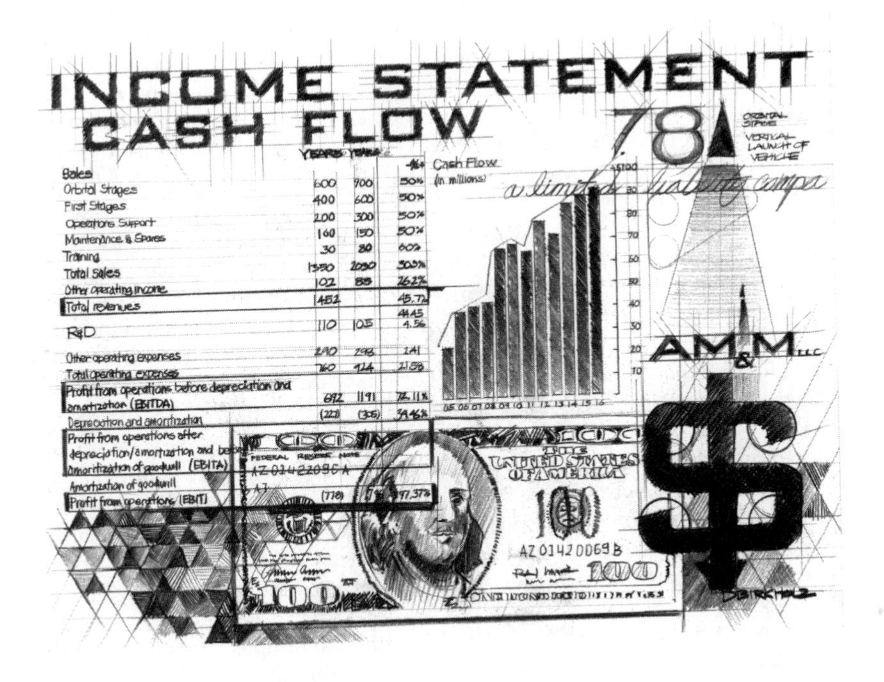

AFTER Tom's presentation, the Group of Seven broke for lunch. It was a beautiful day, and so lunch was served on a covered veranda at the conference center where the Group usually met. The meal was a barbecue, and the main course was brats accompanied by plenty of cold beer—Forsyth was, after all, a Wisconsin boy. A lot of people think billionaires prefer to eat caviar and lobster and drink fine champagne, and that is true for some, especially the ones born to wealth. All but one of this group, however, had come from relatively modest means, and after making their fortunes their tastes had not changed all that much. They were quite content to dine on bratwurst and hamburgers and wash it down with the beer of their choice. After lunch, the Group of Seven and their assistants filed back into the conference room, still engaged in a variety of conversations as the members hashed over what they had heard from Tom.

The next topic would be the financial side of the business plan. It occurred to me that Forsyth might have strategically scheduled this particular talk after a filling lunch and a few beers. As he had explained last week while giving me a preview of the financial picture, this side of the project was a lot harder to validate than the engineering side. For a technical validation of Tom's plan, Forsyth had simply found a couple of eminent aerospace experts who he knew vehemently disagreed with Tom's approach and asked them for an assessment of its technical feasibility. One was an ardent fan of winged space planes, the other an advocate of the big dumb booster approach. Both, however, were very competent engineers, and each had put together a small team of consultants to run Tom's design through the ringer.

The results of their analyses hadn't changed their opinions of how the problem should be tackled, but they had confirmed that Tom's approach was well within the state of the art. Although there had been an occasional request for more detail or backup test data on things like heat-shield design and the parawing deployment loads, in the end, although vehemently disagreeing with the utility and desirability of the approach, they could not fault its basic technical feasibility. Somewhat surprisingly, they even agreed that Tom's development cost estimates were within the ballpark.

On the engineering side, AM&M needed to produce a vehicle with the funding that was available, which in Forsyth's reckoning was from $3 billion up to a maximum of $5 billion. The oft-quoted $50 to 100 billion for a "next-generation launch vehicle" built by the government and the established aerospace industry was so large that it was hard for even the wildest optimists to believe that a market would arise that could ever pay off such an investment. Forsyth and Tom had avoided using classical aerospace costing models, designed to assess the cost of developing new military and government space systems, as they invariably assumed much larger engineering staffs and higher cost structures than would be used by AM&M. Forsyth was confident that there was a reasonable probability that it could be done for $3 billion. But if it ended up costing five—or even seven billion—he felt that the money could be brought together to finish the project, although that would likely result in a write-down for the investors of the first three billion.

He and Tom both felt most comfortable with bottom-up costing, breaking the vehicle down into a list of components and getting specific quotes for those that could be bought "off the shelf" and soliciting rough order of magnitude (ROM) estimates from industry for the other items, based on modifications of existing hardware to qualify them for use on the vehicle. For the vehicle that Tom had conceived, based on technology well within the state of the art, the engineering team had already been able to cost major portions of both stages, including the engines, which would either be existing engines or derivatives of existing engines and could therefore be reliably cost estimated. The cost of the design and development engineering effort was also parametrically estimated, based on a small multiplier times the typical development costs for a business jet or regional commuter jet in approximately the same weight class.

VALIDATING the basic business case was a lot more difficult. Forsyth had broken the problem into two key issues. First, could they find enough buyers for the vehicle, to provide the cash flow necessary to get the company on a sound footing? In the short run, if an affirmative answer could be given to the question, "Is there a market for the vehicle right now?" the venture was likely to develop a positive cash flow relatively quickly. If, of course, Tom came reasonably close to hitting the schedule and cost targets. But these definite programmatic risks could be mitigated by getting the best people on the team, formulating a well-laid-out program, including contingency plans for the inevitable technical snags, and paying close attention to details.

> In the short run, if an affirmative answer could be given to the question, "Is there a market for the vehicle right now?" the venture was likely to develop a positive cash flow relatively quickly.

The second issue was whether or not they would be able to sell enough vehicles to push launch costs low enough to kick off the development of the new markets for launch services that would be necessary to sustain long-term demand for the vehicle. Both problems were tough nuts to crack. But, as Forsyth explained to me, the purpose of a business plan is not to predict the future, but to demonstrate to the investors that serious thought has been given, not just to the technical feasibility of a project, but to the bottom line—in other words, do you really have a good shot at making a profit. To thoroughly ring that out, Forsyth had recently brought Penelope Mundy on board to be AM&M's comptroller; her expertise was in large-scale capital projects. Working with Forsyth, his personal financial manager, and three outside consultants, she had developed the formal business plan that would now be presented to the Group.

The potential markets were the first topic to consider, and Forsyth had of course been studying this subject for years. Early in this project, he had shown me data gleaned from the various FUTRON reports of the last decade or so, analyzing existing markets and predicting future markets for space launch. These and other studies all showed the same thing: the market for launch services was pretty static. Oh, there were ups and downs from year to year, but the trend was definitely flat. Furthermore, studies that investigated the effect that falling costs for launch services would have, while showing some increase in demand for flights, always showed that the total revenues of the launch service providers would actually fall with lower launch costs.

It was this fact that had driven the Group of Seven to conclude that there just wasn't any scenario in which selling launch services made sense. The cash flow would simply be too low to support the investment needed. This in turn led them to the marketing concept of selling launch vehicles, rather than launch services. AM&M would provide buyers with manned access to space, and for an investment of well under half a billion dollars. This approach, although postulating a new market for new launch vehicles, appeared to be feasible—at least on the face of it. After all, it seemed reasonable to anticipate that any nation that was now spending somewhere on the order of $100 million a year on a space program would find a way to buy an AM&M vehicle, thereby gaining the prestige and capability that manned spaceflight represented and breaking up the oligopoly maintained by the United States, Russia, and China.

Forsyth knew that, despite their deep and abiding interest in opening up the space frontier, the Group would need something a little more to go on than the old "if you build them, they will come." Thus, he had hired a group of consultants—retired NASA senior executives, high-ranking military officers, and former U.S. ambassadors—to do a low-key survey of prominent government, scientific, military, and aerospace industry leaders around the globe. Their task was to determine how receptive these leaders might be to the idea of getting their hands on a reusable launch-vehicle system with ownership costs in the neighborhood of $100 million a year. The basic talking points were very "bare bones." There were no pictures of the vehicle, no technical details; Forsyth had not yet obtained an export permit for the design. Frankly, at this point, he did not want a judgment on the feasibility of the particular vehicle, just an anecdotal, off the record, indication of interest. The results of the survey were guardedly optimistic. The idea was met almost universally with interest—and a large measure of skepticism. It was hard for these people to really believe that such a vehicle would ever be available at such a low cost. But this very skepticism was, in Forsyth's mind, an indication that if the vehicle were actually built, it would be seen to have a value beyond its cost.

In addition to the informal survey, which Forsyth realized had a lot in common with reading tea leaves in terms of ability to accurately foretell the future, he and his team had identified 27 nations and institutions worldwide for which the ownership of such a vehicle would support their basic mission. They had reviewed these organizations' budgets and identified adequate levels of discretionary funding that could be used to purchase and operate a vehicle. Forsyth had also brought in actuaries from the launch insurance business to assess the cost of guaranteeing a certain number of successful flights for each vehicle sold. These numbers, together with input from several analysts in the commercial airplane-financing sector, were able to provide relatively solid numbers for planning purposes. It appeared that if flight rates started out at a low rate of two or three flights a year for each vehicle sold the cost of financing the purchase and operation of a launch vehicle and the associated ground facilities could be kept to about $100 million annually. Also, according to the actuaries, the cost for AM&M to provide an insurance policy for vehicle replacement during the first 10 flights, assuming a vehicle overall reliability of 98% and loss of vehicle risk at 1%, could be kept to about $20 million—less than 10% of the vehicle sales price.

ALTHOUGH even all of these data could provide no absolute assurance that vehicles would sell, Forsyth did not expect to have any real difficulty convincing the Group of Seven to go forward. Given sufficient reflection, it should be a relatively easy leap of faith for a group that had jumped into just about every new industry which had come along in the last 30 years, and had usually made money doing it. The long-term market for the space launch industry, however, was more difficult to talk your way around. During the first 25 years of the biotech industry, $100 billion had been invested, and only $60 billion returned. In the early stages of that industry, billions had been raised without any hard evidence that "bioengineering" would ever produce any economically viable products. On the other hand, the market for new and better drugs was well proven. The leap of faith there was that biotech would produce drugs, not that a market for the drugs would arise if they could be developed.

Broadband fiber optics was perhaps a better analogy. With the arrival of the Internet, the telephone and cable companies embarked on a spending spree that wired the globe, based on their belief that there would be immediate demand for broadband into the home. Some $100 billion had been spent to wire the entire planet with little existing market for such services. Broadcast television and cable networks in a sense are broadband into the home, but two-way broadband via the Internet was for years a largely nonexistent market. And when broadband demand finally came, it was too late for many of the companies that took part in the wiring binge. In that instance, a very large investment had been made without any proven market. Forsyth's basis for the AM&M venture was ultimately based on a belief that there would be a future for a new transportation capability. He was convinced that a capability which would provide access to the resources of the solar system had to have a future. It was of course something of a step into the unknown; some of the most vocal proponents for potential space launch markets such as space tourism, extraterrestrial resources, or solar-power satellites could be viewed as wildly optimistic at best, or downright crazy at worst. No, given Kroemer's lemma, Forsyth's sales pitch ultimately had to be based on the assumption that there always had been and always would be a future for cheaper transportation.

The Group settled back into their chairs, and Penny Mundy got up to begin her presentation. The outline of the business plan was simple. A $3 billion investment spread over four years. An arithmetic increase in the number of launch vehicles sold, starting with 4, then 6, 8, 10 per year, and so on. Launch costs would be starting at $5000 per pound, and would be decreasing by half every 18 months for the first seven years, until the price reached $200 per pound. Average launch rates for the vehicles would be at twice per year initially, building to an average annual flight rate of 25 by year 16.

Ms Mundy summarized the basic business concept with a viewgraph showing a linear growth in vehicle sales, superimposed on an exponentially growing market for launch services. The key point was that it would take a long time for launch services to amount to a substantial market. By year 16, 12 years after production of the vehicle was initiated, the launch services revenues were shown to barely surpass those from launch-vehicle sales. Penny's next viewgraph showed 16 years' worth of pro forma income statements, based on this and the other assumptions she had developed with Forsyth. The pro forma statements that she was presenting were not so much a prediction of the future, but rather an exercise in analyzing and evaluating the reasonableness of the basic assumptions and expectations of the Group. Certainly, as business plans go, it was not wildly optimistic, with cash flow only going positive in year 6 and reaching net payback in year 11.

The key point was that it would take a long time for launch services to amount to a substantial market.

Forsyth had explained to me the pitfalls of the standard MBA approach to devising and analyzing business plans. The MBA approach starts by tallying up all of the projected expenses and incomes, along with the probabilities of certain outcomes. These results are then compared with and evaluated against other uses of the same investment funds. Internal rate of return and net present value, or future value, are concepts that were developed as a means to allow large corporations to evaluate different capital or budget proposals against each other, providing a way to rationalize the decision-making process. However, even when used in the situations for which it was originally developed—big business or governmental institutions—it has to be applied with some common sense and foresight. Sometimes a seemingly poor investment is necessary to stay competitive, or just to remain a player, in a particular industry. Sometimes a seemingly rock-solid investment is based on a static model of the world that is just plain unlikely to persist. Certainly the history of business and government shows mixed results as far as the ability of these methods to predict which investments are going to be wildly successful or dismal failures.

As she worked her way through the pro formas, Penny was careful to underscore the basis for the figures that were used. The number of vehicles sold each year, for example, was based on the survey of potential customers. Vehicle flight rates were based on engineering estimates and Monte Carlo simulations. Then there were those factors that would not be under AM&M's control, such as whether or not launch costs would indeed fall by half every 18 months. Forsyth intended to loudly proclaim this as "Forsyth's Law," but whether it would become a self-fulfilling prophecy as Moore's Law had was certainly open to question. The development costs, of course, were based on careful estimation of both the hardware and the engineering effort that would be required.

As Penny went through the line-by-line breakdown of the development cost estimates, Number 5 raised an objection, as Forsyth had predicted he would, to the pay structure for the engineers. Number 5, who had made his fortune in creating and trading in new investment vehicles and thus had never had the experience of managing a large organization, voiced his opposition to the concept of paying engineers two to three times the market rates. Why pay them with scarce cash early in the venture, when it is standard practice for new technology companies to pay in stock options, thereby enhancing cash flow in the early years and paying with only relatively minor stockholder dilution in later years, he wanted to know. Penny responded with three reasons why the proposed pay structure would be most efficacious for attracting and compensating the talent necessary to see this project through. First, she noted that in the aerospace industry, stock options—except for senior management—are not the norm, and in fact are quite rare. Therefore, experience with and expectation of stock options is not part of the aerospace engineers' culture, and they are unlikely to be a motivating force for recruiting the sort of people AM&M needed to employ. Second, given AM&M's large capital structure, would the investors really be willing to trade what might be 30% of the future equity value of the company for a 3 or 4% savings in total up-front expenses, from reduced payroll costs? Last, there were the negative side effects that making millionaires out of your employees could have if the company eventually went public. Aerospace has a very long product cycle, often measured in decades, and stock options have a way of motivating early retirement, which would disrupt the "corporate memory" and be detrimental to ongoing developments and product improvements. There is also the internal discord and resentment that can develop with stock options, as later hires watch early hires become millionaires while they don't. These organizational and morale issues could easily degrade AM&M's future performance and competitiveness, especially because maintaining a knowledge base for systems designed 10, 15, or 20 years ago would likely be vitally important. Forsyth and Penny's preparation on these issues seemed to have paid off. Number 5 settled back into his chair, apparently satisfied.

Penny finished up with a short discussion of the economics of the launch facilities development business. Forsyth was already shopping around, holding talks with several of the larger global construction management firms. The plan didn't make the facilities look like a terribly good investment, but this analysis really only dealt with the bare-bones launch facilities. Penny explained that

Forsyth's pitch to the construction firms was based on exclusivity, which should provide a substantial competitive advantage in obtaining work on related facilities. The total investment here was also small enough that it could practically be justified as a marketing tool—a prestige project with high visibility in practically every major participant in the global economy.

After Penny sat down, Forsyth got up to tie together all that had been presented. He reminded the Group of what had brought them together in the first place—not merely a quest for the most profitable investment opportunity that was available, although you couldn't say with certainty that this wouldn't turn out to be it. No, their motivation was to take a shot at opening the space frontier with as high a probability of success—and profit—as possible.

"I know most of you have invested in the space industry before, turning a large fortune into a small one. But this time we've got the resources, we've got a marketing plan that make sense, we have a vehicle that makes sense, and we have the engineering team that can get it done. There are risks involved, but I've laid all my cards on the table, madam and gentlemen." He paused. "I am putting up $500 million for the first round."

After Forsyth's speech, there was a dinner that was considerably lighter than lunch had been. Following that, Forsyth negotiated with the other six members of the Group one on one. By the early hours of the next morning, he had gotten commitments totaling $500 million, in addition to the $500 million that he was putting up himself. It took another month for the lawyers to make it all formal, but there it was—a billion dollars committed to the project. Forsyth expected that this would see them through the flight-test program, which would validate the concept. They now had about 24 months of breathing room before the next round of financing—to put the vehicle into production—would be required.

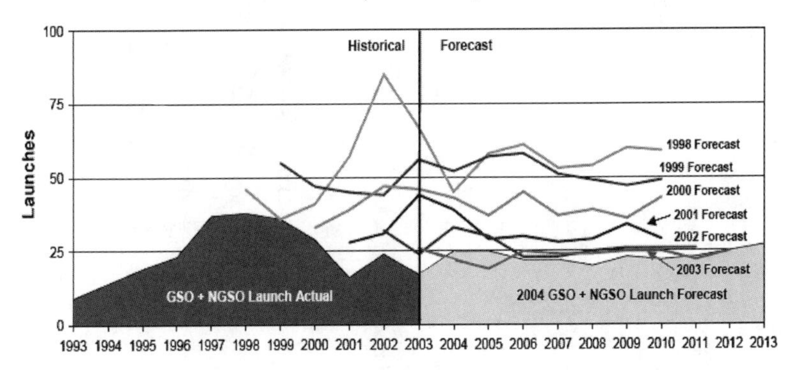

* 2004 Commercial Space Transportation Forecasts, May 2004, FAA/AST.

AM&M Pro Forma Statement

Year of Business Plan	0	1	2	3	4	5	6	7	8
System Sales Price ($M)						250	250	230	220
First Stage Sales Price($M)						75	75	70	70
Orbital Stage Sales Price($M)						175	175	160	150
Number of DH-1 first stages sold						4	6	8	10
Number of DH-1 orbital stages sold						4	6	8	10
Total Vehicle Sales ($M)						1000	1500	1840	2200
Cost of Goods Sold ($M)						540	810	992	1180
Gross Profit ($M)						460	690	848	1020
R&D expenses ($M)	50	250	750	1000	1000	400	300	400	500
SG&A expenses ($M)						60	90	110.4	132
Net Income Before Taxes ($M)	–50	–250	–750	–1000	–1000	0	390	448	520
Cash Out ($M)	–50	–250	–750	–1000	–1000	–1000	–1200	–1502.4	–1812
Cash In ($M)	0	0	0	0	0	1000	1500	1840	2200
Net Cumulative Investment ($M)	–50	–300	–1050	–2050	–3050	–3050	–2750	–2412.4	–2024.4
Total no. of orbital stages produced						4	10	18	28
Fight rate per orbital stage per year						2	2	3	5
Total flights per year						8	20	54	140
Tons to orbit per year						20	50	135	350
$/LB to LEO, falling 1/2 every 18 mo.						2000	1260	793.8	500.094
Revenue from launch services ($M)						80.00	126.00	214.33	350.07

Year of Business Plan	9	10	11	12	13	14	15	16
System Sales Price ($M)	210	200	190	185	177	170	160	140
First Stage Sales Price($M)	70	70	65	65	60	60	60	50
Orbital Stage Sales Price($M)	140	130	125	120	117	110	100	90
Number of DH-1 first stages sold	12	12	12	12	12	14	16	18
Number of DH-1 orbital stages sold	12	15	18	20	23	26	30	36
Total Vehicle Sales ($M)	2520	2790	3030	3180	3411	3700	3960	4140
Cost of Goods Sold ($M)	1344	1506	1662	1752	1902.6	2052	2184	2304
Gross Profit ($M)	1176	1284	1368	1428	1508.4	1648	1776	1836
R&D expenses ($M)	500	500	550	600	650	700	700	750
SG&A expenses ($M)	151.2	167.4	181.8	190.8	204.66	222	237.6	248.4
Net Income Before Taxes ($M)	676	784	818	828	858.4	948	1076	1086
Cash Out ($M)	–1995.2	–2173.4	–2393.8	–2542.8	–2757.26	–2974	–3121.6	–3302.4
Cash In ($M)	2520	2790	3030	3180	3411	3700	3960	4140
Net Cumulative Investment ($M)	–1499.6	–883	–246.8	390.4	1044.14	1770.14	2608.54	3446.14
Total no. of orbital stages produced	40	55	73	93	116	142	172	208
Fight rate per orbital stage per year	8	11	14	16	18	20	22	25
Total flights per year	320	605	1022	1488	2088	2840	3784	5200
Tons to orbit per year	800	1512.5	2555	3720	5220	7100	9460	13000
$/LB to LEO, falling 1/2 every 18 mo.	315.0592	225	200	200	200	200	200	200
Revenue from launch services ($M)	504.09	680.63	1022.00	1488.00	2088.00	2840.00	3784.00	5200.00

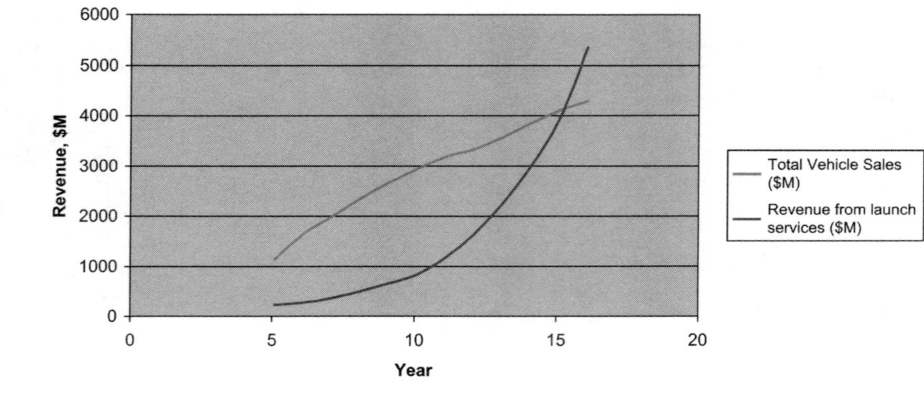

CHAPTER 9
Mazes, Stop Cords, and Skunk Workers

IN the year since Tom had joined AM&M, he had increased his team from 50 engineers to 150, with about 50 support personnel. Right from the start, Tom had laid out an unusual system of compensation for engineers. There were only three pay grades. The first, trainee level, paid pretty much the market wage for graduate engineers. The second, engineer level, paid about 150% of the average wage for engineers in their peak productive years, or approximately three times the starting wage. The third level, vehicle engineer, paid about twice the second, or six times the starting wage. In this level Tom included himself. Although the top bracket was primarily meant for senior managers, there were no hard lines, and some of Tom's engineers, though only in their late 20s, were getting the third-tier wage.

Tom's system was designed to accomplish three objectives. First, his people were going to be working long hours, and he didn't want them worrying about financial matters on the side. Second, a lot of the negative workplace politics that Tom had experienced in engineering organizations was based on differences in compensation. Too much management time was taken up with performance reviews and dealing with hurt feelings over raises that were given or not given. Under Tom's system, everyone in the junior and middle levels was getting basically the same pay. If they were taking on unusual responsibilities and making proportionally larger contributions to the project, that contribution was recognized by advancement to the upper tier of income. The third reason was to attract and retain the best and brightest. Forsyth and Tom were agreed that, with the heavy capitalization required for the development and production of a reusable launch vehicle, high-flying stocks were by no means a certainty. The large capitalization needs meant that most of the equity ended up with the investors, not the employees.

Like some of the more innovative companies at the turn of the millennium, titles were downplayed, and the organization was kept extremely flat. This was facilitated by the three-tier system, where you could make top dollar in any job, if you were contributing enough. However, the company did have a dress code. Each of the three compensation levels—trainee, engineer, and vehicle engineer—had slightly different dress codes imposed, so that, together with identification badges, within the broad categories your status was evident, and small courtesies were shown to the vehicle engineers. Thus performance and contribution were motivated by money and status as starting points.

TOM also worked hard to instill a strong, universal embrace of the corporate culture—and the vehicle design itself. He wanted everyone to share a feeling of being part of a very unique family that was doing something really important. That way, security was naturally tight, and the company presented a uniform face to the outside world. Forsyth had been adamant about the importance of two key political, or organizational, issues. One was basic honesty and integrity. No one would buy something as risky and as unreliable as a launch vehicle from a company or people who were not trustworthy. If the rest of the world was going to trust AM&M, then the employees all had to trust each other.

The other was public perception of the company. As John Forsyth and Tom were well aware, every engineering project of any complexity is like running a maze where you aren't sure if there is a way through. As each engineering decision is pursued through design, construction, and testing, problems—often seemingly insurmountable—will appear. The first impulse of the inexperienced engineer is to trade the known problems of this design approach for the unknown problems of some other approach. But, "this way madness," as the saying goes, which soon leads to more madness, and—if left unchecked—to financial ruin. Every seemingly blind passageway has to be thoroughly explored, and sometimes a wall has to be knocked down. Sometimes retreat is necessary, but never too far, as you make your way slowly through the maze to the design objective.

Every engineering project of any complexity is like running a maze where you aren't sure if there is a way through.

The problem with this process, especially for a startup that is totally dependent on one new product, is that it can be pretty terrifying as you go through it—one difficult or impossible challenge after another has to be conquered, or you die. If these "little" problems along the way lead to rumors, wild speculation, disgruntlement, and second guessing among the engineers, then it is quite likely that some of that will leak out. Especially in this age of email and the Internet, such things can spread quickly. The next thing you know, there is a general perception among the public, but more importantly among potential customers and investors, that the team is following the wrong approach, they don't know what they're doing, they're wasting money, or whatever. Such rumors and innuendoes can be disastrous where large amounts of capital have to be raised over a period of years. By building a tightly knit organization where all team members are well informed and have faith and confidence in the overall approach, in their coworkers, and in their management, that confidence and optimism—that sureness—can be presented to the outside world, and especially to the investors.

Engineers often disagree about what approach is best, which approaches are possible, and which can be accomplished with the money and time available. Tom and John Forsyth were agreed that you could rarely convince someone by argument. People can have their minds changed by education, yes, but more useful—especially when time is of the essence—is authority, in the form of someone they look to with trust and confidence, someone to whom they will defer judgment. But authority is a two-way street—management also needs to trust the engineers and defer to their judgment within their discipline or area of expertise.

The Space Shuttle *Challenger* loss illustrates the perils of ignoring this wisdom. There are many perspectives from which to view that tragedy, but in retrospect it seems clear that launching in cold weather posed a serious risk. The engineers most familiar with the solid rocket booster and the O-ring design were unanimous in this key opinion. The *Columbia* loss was similar. After launch the most knowledgeable engineers were concerned for the safety of the shuttle because of the insulation striking the orbiter. But management again remained passive. So, what was the organizational failure? Some argued that it was a failure of presentation, that if the data which existed had been presented in a clearer, more cogent way, management would have responded, and postponed the launch. Or, in the case of *Columbia*, they would have ordered an inspection of the wing by the astronauts. Others argued that it was a failure of a mindset, one that required proof that conditions were unsafe rather than proof that they were safe. Others placed the blame on a conflict of interest—the pressure on management to launch more flights pitted against the desire of the engineers in the trenches to make sure that nothing ever blew up.

There was undoubtedly a bit of truth in all of these explanations, but Tom was convinced—and Forsyth concurred—that the missing factor was what Tom called the "Toyota factor." In a Toyota production facility, anyone—from the lowliest assembly worker to the manager of the plant—can pull a cord and stop the entire production line if they see a problem. And the line doesn't start again until everyone is satisfied. The key here is the authority given even to the most junior assembler to stop the line.

This was the way it was at AM&M. The authority to stop everything was given to everyone from the trainees up to the chief engineer. And with that authority came a true open-door policy—if the most junior engineer wanted to talk to the most senior manager, the manager had better take the time to listen, or everything could come to a halt.

The engineers were empowered in other ways, chiefly through budget and purchasing authority. Each was given their own budget—usually negotiated—for the part of the project for which they were responsible. They were told to spend the money as though it was their own. They had access to experts in the legal aspects of contracting, negotiating, and purchasing, to use or not as they chose. Enough training was provided to keep them out of trouble, so that they would know when they were actually committing funds and when they were not. They could spend their money without oversight and signed all of their own purchase orders. Each had a company credit card and a company Pay Pal account that they could use at their discretion. Controls were all retrospective. Once a month—more often if the charges were unusual—they were called in by procurement supervisors. These procurement supervisors—all older women as it turned out—reminded the engineers of those junior high school instructors who never cracked a smile and were all business. They went over the expenses with a fine-tooth comb, sometimes going back over several months, or even back to the beginning of the project. But the procurement supervisors did not have the authority to spend or to prevent spending. Their job was simply to call the engineer's attention to what he was doing and to make sure he was thinking it through. It was the engineer's job to get his project task completed or his subsystem to work, preferably on time and under budget.

As real funding kicked in, Tom went to a six-day schedule, and most of the engineers found themselves putting in 10-, 12-, or even 14-hour days. At the same time, the company provided support for the families. Personal assistants, who could arrange for home repairs, sign a child up for school, and even purchase major items were made available. The company was generous as well with psychiatric benefits for the entire family, if problems arose. More importantly, the company encouraged spiritual formation. In this they were aided by what many people said was another great awakening, a religious fervor that was sweeping the nation, and, what with the recent resurgence of the Roman Catholic Church, had a strong Catholic flavor. Forsyth was himself a nominal agnostic, but Tom was somewhat more serious about his religion. Both agreed, however, that a religious influence made for a happier work force. Their home lives were happier, and they were less bound up with their own egos; they simply had a better, healthier sense of their place in the world.

Company policies followed Ross Perot's dictum: "if your wife can't trust you, how can we?" The idea was to avoid as much emotional turmoil as possible—in the context of what was, at least for the next five years, a very demanding job—by encouraging dedication to family. The goal of these policies was to create, within a year or two, the close-knit team environment such as had been achieved by organizations like the Skunk Works. To Forsyth, it appeared that Tom was doing a better job than anyone could have expected. He was quite encouraged to see that they seemed to be successfully clearing the first major hurdle: building an organization that would bring out the best in everyone.

The goal of these policies was to create, within a year or two, the close-knit team environment such as had been achieved by organizations like the Skunk Works.

ANOTHER aspect of their approach to developing a cohesive and well-integrated workforce, which at the same time provided a certain formality, was to follow the old IBM approach. That is, when selling something that is perceived to be risky and perhaps unreliable, it is important that the entire organization—sales force, engineering team, service people—come across as staid and conservative. IBM might never have had the best computer, but they had the most trusted support structure, and so dominated the market. Tom was adamant that they were going to have not only a working reusable launch vehicle for sale, but one in which the marketplace would have complete confidence.

Eventually, if they were successful, competitors would rise up to challenge AM&M, and even though it seemed likely that they would have a 10-year head start on the competition Forsyth's strategy was to create an organization that would stand behind its vehicles. The same organization, which was going to have to sell the vehicles to very conservative, risk-adverse organizations—NASA, the Air Force, even the Army and Navy, if he had his way, as well as foreign allies—was also going to have to sell the vehicle to the regulatory agencies.

John Forsyth was by no means blind to the regulatory challenges they faced. Just getting permission to fly a single vehicle had been a contributing factor to the demise of more than one startup rocket company. But he was also aware that once a regulatory regime was successfully negotiated by—and to a large extent, created by—the first company in a new industry the advantages that they thereby gained would be a major impediment to competitors following behind.

Forsyth's strategy for tackling the regulators had two primary objectives. The first was to influence the Federal Aviation Administration's Associate Administrator for Commercial Space Transportation, who had been given the authority to regulate space launches and vehicles in 1995, in order to get a set of workable and tolerable requirements for certification of a manned, reusable launch vehicle. Getting the vehicle certified in the United States would be the first step to worldwide acceptance of such basic operational issues as overflight of populated areas.

The second objective was to get permission to export the vehicle. To meet their sales objectives, they were going to have to be able to sell vehicles to foreign customers. And in some ways, this seemed a more daunting problem, requiring the approval of the State Department, under the International Traffic in Arms Regulations (ITAR). After all, wouldn't you essentially be selling an intercontinental ballistic missile? Forsyth's ready response to that question was to point out that the only serious attack on the continental United States in the last 100 years had been made using U.S.-owned airliners. Clearly, a fully loaded Boeing 747 was more dangerous than the vehicle they were going to build. Besides, with the deployment of the first phases of a Star Wars defense system it was arguably easier to shoot down than an airliner. But he knew it was still going to take a lot of work before the ITAR and technology transfer issues could be dealt with in the same way they were for commercial airplanes.

Over the next two weeks, Tom met with his senior people to work out a plan for dividing the engineers into project teams. Tom was a believer in Augustine's Law, which said that, as you add more people to a project, at some point the amount of work they get done goes down because the additional people are less productive and the time consumed in communication and coordination with the new people dissipates more effort than they add. This was the delegation tax writ large. One of Tom's criteria was to keep each team to 50 or fewer engineers.

The project was broken up along vehicle functional and structural lines. One team was given the prototype orbital stage, a second team the prototype first stage with rocket-powered landing, and a third, smaller, team an alternative first-stage prototype using jet engines for landing. The fourth team was assigned the production orbital stage, and the fifth team to the production first stage. Later on, manufacturing teams for the first and second stages would also have to be organized. A flight-test team for the first stage, and subsequently for the integrated first- and second-stage flight tests, would be needed. There would also be a manned systems team that would work with a small corps of test pilots to design all of the crew systems and equipment, from the ejection seats to the space suits, as well as a flight simulator. They would also have the honor of tackling that perennial problem of how to go to the bathroom in space.

Of course there would be a lot of movement between the teams. Obviously many of the key design people would start out on the prototype and then move on to the production vehicle. The pay structure and the flat organization chart was certainly going to facilitate that movement. People didn't have to worry about whether or not their boss was going to appreciate what they were doing if they were sent to another team because, aside from the cost-of-living adjustment, the pay was going to remain high and stable. Unless, of course, it went higher still if you performed very well and were made a vehicle engineer.

AM&M Organization Chart

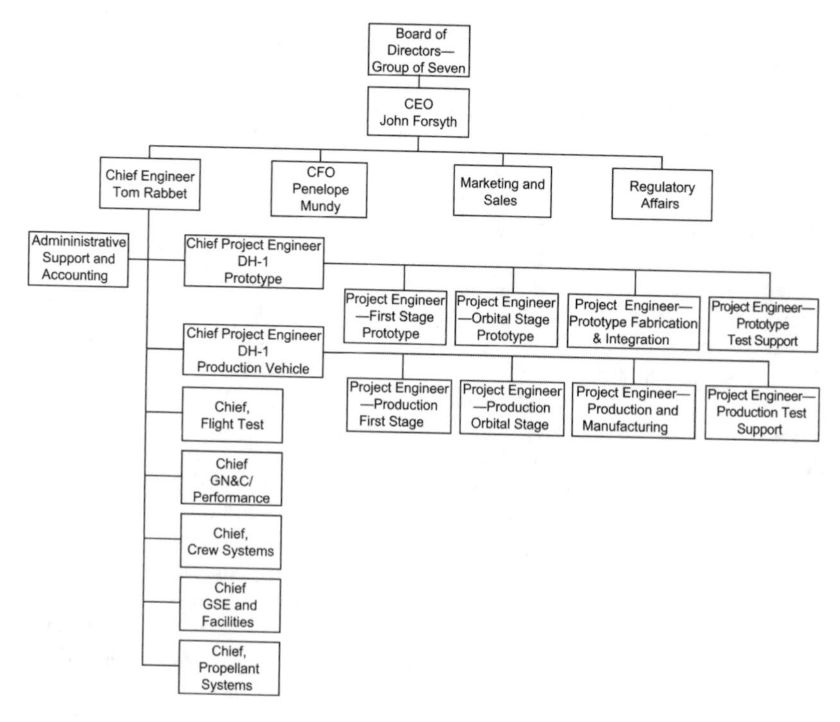

CHAPTER 10
Fuel Tanks, Heat Shields, and Fire Walls

ONE area of intense activity involved detailed design of the airframe for the orbital stage, consisting primarily of the propellant tanks and thermal protection systems. I asked Tom to give me an overview of this effort. As part of their earlier design exercises, Tom and his team had considered the various functions that the parts of the vehicle structure need to provide, including possible ways of arranging or combining those functions to arrive at an optimal solution. In this, he borrowed a technique that he had seen applied to the design of cryogenic tanks for ocean transport of liquefied natural gas.

The most complicated part of the vehicle structure would be the orbital-stage liquid-hydrogen tank, which would have to deal with the very low temperature of the fuel on the inside, as well as the very high temperatures caused by reentry heating on the outside. So this tank was the starting point. Its first function is that of containment. Hydrogen gas is notorious for its ability to leak through extremely small pores or cracks. Any such leakage presents a potential fire hazard while the vehicle is in the atmosphere. The second function is to support the propellant pressure loads that are required to feed the turbopumps. For the hydrogen tank, this means 20 to 30 psi over atmospheric pressure, growing to an absolute pressure—once the vehicle leaves the atmosphere—of 30 to 45 psi. The third function is also structural: the tank needs to support the payload bay and the oxygen tank above it as well as take up and distribute the thrust load from the engines. The fourth function is to resist the so-called ground-handling loads; these are the loads imposed on the vehicle while it is being processed, stacked, and readied for flight. The fifth function that the hydrogen tank needs to provide is cryogenic insulation. Liquid hydrogen boils at a temperature of −420°F, which is more than 100° below the liquefaction temperature of air. If any air started to liquefy on the outer surface of the vehicle, it would present two serious problems. First, rapid heat transfer to the condensed air would cause the hydrogen inside the tank to boil off quickly, and second, there would be an explosive hazard should the liquid air come into contact with any hydrocarbon substances.

The sixth function that the tank structure needs to provide is a long fatigue life, meaning the material used must be resistant to the initiation and growth of small cracks. This implies a strong, tough, but not brittle material that would leak before explosive rupture. The seventh function is protection from the high temperatures and heat loads of reentry. To make this job easier, the exterior would preferably have a noncatalyzing surface—in other words, a surface that does not facilitate the recombination of the ionized gases which surround the vehicle during reentry. Recombination dramatically increases the heat load on the surface. A related function, number eight, is that of a heat sink—there has to be some place for the heat that does come through to go, so that no part gets too hot. Finally, the ninth function that the vehicle structure must provide is thermal management in the vacuum and alternating shadow and sunlight of low Earth orbit.

One design approach, popular in recent years and the subject of a good deal of development effort during the National Aerospace Plane (NASP) and (to a lesser extent) the X-33 programs was to separate the various structural functions. An aluminum tank (or in some concepts, a honeycomb composite tank) designed to contain the propellants would be placed inside a separate, exterior aeroshell structure. These designs typically ended up having one structure for carrying the vehicle thrust loads, another set of structures to stiffen the tank walls and to support their internal pressure, a system of cryogenic insulation mounted on the outside of the tanks, a high-temperature reentry structure to take the aerothermal loads, additional structure for distributing the loads to the vehicle tanks, and finally an external thermal protection system. Not surprisingly, these concepts were uniformly heavier than was acceptable, often weighing as much as 4–5 pound per square foot. And because the underlying structure comprised so many disparate pieces, there was no good heat sink to absorb the reentry heat loads, and the structure that supported the thermal protection system tended to be driven to very high temperatures.

TOM knew that the success of his vehicle hinged on an efficient, lightweight structure that combined as many of these functions as possible. The lowest-weight structure that had been devised so far for a reusable vehicle was generally considered to be the one designed during the Air Force HAVE REGION program in the 1980s. This concept used a honeycomb sandwich structure consisting of an outer layer of Rene 41, a honeycomb core of Titanium 6Al-4V, and an inner skin of Inconel 718. The Rene 41 (a nickel-based superalloy) could in theory withstand the reentry heating by itself—if a sufficiently gentle, low-heating-rate gliding reentry was used and if the entire structure was allowed to reach relatively high temperatures. This is a so-called "hot structure," in which the airframe is made from high-temperature materials that can be allowed to and are designed to get very hot. The honeycomb provided the cryogenic insulation for the propellants and the structural stiffness for ground handling. Propellant tanks with specific weights as low as a half a pound per cubic foot were designed, and sample pieces of structure were fabricated. The weight per unit area of this structure for upper, cooler surfaces was estimated to be in the range of 1.5–2.5 pounds per square foot. For the hottest surfaces on the underside of the vehicle, the range was 3–4 pounds per cubic foot.

When all was said and done, however, this structure presented several difficulties. First, it was extremely difficult to fabricate large tanks using the high-temperature vacuum brazing process that was required. Small pieces of structure can be made by stamping the inner and outer metal layers to shape, milling the titanium honeycomb core to shape, spreading a braze alloy paste on the upper and lower sheets, sandwiching it all together, and placing the entire structure in a thin stainless steel bag on which a vacuum is drawn to force the sheets into intimate contact with the honeycomb. It is then placed in an oven, brought up to the brazing temperature, cooled, and the stainless steel vacuum bag cut away—not very easy for a big tank. The second problem had shown up in a similar structure used on the B-58 Hustler supersonic bomber. Tom had once talked to an engineer who had worked on that plane, which used stainless steel skins and honeycomb panels for the wings and fuselage. This did produce a high-strength, lightweight, high-temperature airframe. However, the structure had a relatively low fatigue life because of the thin sheet gauges used and the possibility of contaminates getting into the honeycomb between the sheets and attacking the braze alloy. And it too was very difficult to manufacture.

The HAVE REGION structure divided the skin thickness needed for strength between the outer and inner skins, meaning that the skin was everywhere very thin, which increased the likelihood of through-cracking. If the outer skin cracked, ingress of moisture and consequent corrosion were a problem. If the inner skin cracked, a thermal short circuit was created, and there would also likely be problems with delamination, as any liquid or gas trapped between the sheets would expand as a result of reentry heating.

For a ballistic reentry vehicle, the heating rate on the base of the vehicle would be sufficiently high that additional heat shielding would be required even if a HAVE REGION-type structure were used. But the sides of the vehicle, which were not directly exposed to the most severe reentry flow, would actually see relatively low heat-flux rates, and so a hot structure was not required. Running a hot tank would present another set of problems. The residual propellants in the tank would necessarily have to be vented during reentry, and upon reaching the lower atmosphere where the tanks would cool you might end up with an internal pressure below ambient, resulting in compressive, or crushing, forces on the tank. And, if the tank were vented at all times, you couldn't use internal pressure to stiffen the tank in order to improve its ability to sustain reentry or landing loads;

that was one technique that Tom was especially interested in as a way to build the lightest possible tanks.

The Lockheed X-Rocket study of the late 1980s, which had also baselined honeycomb structures, highlighted yet another difficulty of this approach when you tried to build tanks for liquid hydrogen. The inner face of the honeycomb tank was necessarily at the temperature of the liquid hydrogen, while the outer surface was at ambient, providing a nearly 500°F temperature gradient. This meant that the inner sheet wanted to contract—a lot—because of the very low temperature, whereas the outer sheet wanted to stay the same. This created high thermal stresses. On reentry, the problem was made even worse by the rise in temperature of the outer skin, which would then tend to expand, creating even higher thermal stresses. Honeycomb structures also did not provide a very good heat sink behind a heat shield because most of the structure was not in good thermal contact with the shield. This meant that the heat-shield backface temperature, where it would presumably be bonded or joined to the tank wall, would necessarily be high.

The bottom line for honeycomb structures was that, although they could be lightweight and might prove valuable in some portion of a vehicle, the technology for fabricating large, reliable, reusable tanks was not in hand. The prediction of the fatigue life of such structures was difficult, and, although it seemed at first that they would be very light, it was all too likely that as the various design problems were resolved the weight would grow. Honeycomb structures do not bear concentrated loads very well, and reinforcements add weight. Composites have this same problem.

Tom's approach of combining as many of the vehicle's structural functions in as few structures as possible led him to settle on a unitary aluminum structure, which naturally made for an effective heat sink. By using a single-wall aluminum tank to contain the fluids, support the flight loads, and take the handling loads, he got an efficient, lightweight structure that could take a lot of heat. Placing all of the structural functions in a single tank wall had other advantages: with all of the aluminum in a unitary structure, you achieved greater stiffness, lower stress levels, and greater resistance to fatigue cracking. All that was needed to complete the hydrogen tank was suitable internal cryogenic insulation and a suitable external high-temperature thermal protection system.

By using a single-wall aluminum tank to contain the fluids, support the flight loads, and take the handling loads, he got an efficient, lightweight structure that could take a lot of heat.

Tom's perspective was that cryogenic insulation just wasn't the big problem it was usually made out to be. After all, cryogenic liquefied natural gas was shipped all over the world in large tankers, and the nation's highways were constantly traversed by tanks of liquid hydrogen, liquid nitrogen, and liquid oxygen. Cryogenic fluids were stored at and used by hospitals and industrial facilities around the world. So, clearly cryogenic insulation was a problem that industry knew how to handle.

BUT despite endless technology development programs, leading up to and continuing even after the introduction of the space shuttle, there had been little practical development in this area for launch vehicles. Part of the reason was that it really didn't apply to expendable vehicles, and there was still only one reusable vehicle in the world—the space shuttle, which didn't have reusable cryogenic insulation on the external tank. The expendable external tank used polyurethane foam sprayed onto the outside of the tank, which worked all right, except when it delaminated and damaged the orbiter during liftoff, as happened to the Shuttle *Columbia* in 2003. The Saturn S-IVB stage had used internal insulation, which proved to be durable enough for a single flight. But Tom also remembered talking to a Pratt & Whitney engineer who had told of an incident during the Saturn 1B program. The insulation in its hydrogen tank had been found to be breaking away, and there was concern about how much could break away before it affected the engines. This engineer took a bucket of the stuff and ran it through an RL10 engine on the test stand; the engine just ground it up and burned it in the chamber with no problem. Whereas that spoke well for the robustness of the RL10, it was also an indication that the internal insulation problem really hadn't been solved in the Apollo days. But getting it right was crucial to the success of a completely reusable launch vehicle.

Fortunately, the problem could be attacked on a relatively small scale. All you really needed was equipment for storing and handling liquid hydrogen and an aluminum bucket. Then you simply tried various ways of lining the bucket and measured the heat-transfer rates. Tom had been using an outside laboratory to test various foam materials and bonding techniques. One material—solomide, developed for low-flammability applications in airliners—was of particular interest. If bonded to the tank walls with an adhesive designed for cryogenic use, it looked very promising. One such cryogenic adhesive was the famous "blue death" used by Lockheed; it was so named because it was dyed blue to help technicians be sure that they had a uniform application of the stuff, and if it stuck to something that you didn't want it to, that something was dead. Solomide, which is a Kapton-polyamide foam available from the Boyd Corporation, is available as a closed-cell or a porous foam. At first blush, it seemed that the closed-cell type would be the most effective. However, the closed-cell foam suffered a serious reduction of insulating capability as a result of contraction of the blowing gas (used to form the foam) on exposure to the liquid hydrogen. The porous, open-cell material, on the other hand, allowed gaseous hydrogen to penetrate the foam and thus was very effective. The density of the material was 0.6 pounds per cubic foot, and a 2-inch layer on the inside of the hydrogen tank added a mere 0.15 pounds per square foot when the bonding compound was included.

The other problem—that of a reentry heat shield or thermal protection system—had always been over-dramatized by NASA, in Tom's opinion. The old NASA promotional pieces would tell you that "nothing known to man can survive the heat of reentry," and then proceed to describe, in detail, how the Apollo reentry heat shield had been constructed. They took a honeycomb sheet, attached it to the base of the vehicle, and filled each hole, one by one, with what was essentially bathroom caulk (phenolmethylsiloxane). They then carefully x-rayed the heat shield to see that every hole was filled and free of voids. If they found any that weren't completely filled, technicians would drill it out with a dentist's drill and refill it. The extraordinary pains that were taken to make sure each hole was filled made the heat shield seem high tech to some extent, even though it was actually quite simple. During reentry, the caulk would ablate—it would vaporize, and the vaporized material would carry the heat away. On the earlier Mercury capsules, a simpler, albeit heavier, solution had been used. Here the heat shield was just a large chunk of beryllium metal that absorbed the heat. It was heavy, but effective. A pound of beryllium, with a specific heat of 0.2, can be raised to 1000°F without losing its strength; that comes out to 200 Btu per pound. For a typical reentry heat load of 2500 Btu per square foot, a heat shield consisting of 10–25 pounds of beryllium per square can survive the "impossibly high temperatures of reentry" merely by absorbing the heat load.

The Chinese had developed another novel but usable low-tech solution. They glued wooden blocks, appropriately contoured, with the end grain facing the reentry airstream. The wooden heat shield would char and ablate during reentry, just like the caulk material on the Apollo capsules. The fact that you could build a serviceable heat shield for reentry from space out of wood certainly showed that the basic problem was not insurmountably difficult, and so Tom had always regarded this too as a rather straightforward challenge. Nonetheless, for the orbital stage he needed a thermal protection system that would be highly reusable and lightweight. Here the team made proper use of the trade study—not to compare dozens of ill-defined concepts in an analytical process that would supposedly determine which concept was the "winner," but rather to systematically consider all of the options for a particular design problem, so that when an engineering judgment was finally made it was as well informed as possible.

TOM was interested in rethinking the whole reentry protection problem. He took some time to elaborate on something that has been too often overlooked in the design of heat shields: the fact that they must perform three functions. First, the outer surface must be capable of withstanding the high temperatures generated during reentry; second, there must be insulation to keep the interior of the vehicle from seeing those high temperatures; and finally, there must be a heat sink that will absorb the heat which does flow through. Shuttle tile material, for instance, reradiates about 98% of the heat load applied to it, which means that approximately 2% will flow through the tiles to the underlying structure. Heat flow is not the same as temperature, of course. Heat flow is caused by a difference in temperature, with heat flowing from the high temperature to the low temperature, ultimately raising the low temperature. Insulation provides resistance to that flow of heat. The so-called R value for home insulation is a measure of the amount of heat that will flow through a certain thickness of insulation for a given temperature differential. An R factor of 40, for example, means that 1/40 of a Btu per hour travels through each square foot of the insulation for each degree of temperature difference between the warm side and the cold side.

In summertime, as heat leaks in from the outside, an air conditioner can be used to continually remove excess heat from the interior of the house. In the space shuttle, however, there is no air conditioner. Remember, only 98% of the heat load applied to the tiles will be reradiated to space; 2% still makes it through. So if a total thermal load of 2500 Btu is applied per square foot of surface area, that means that approximately 50 Btu per square feet will pass through the insulation. Fifty Btu will raise the temperature of the room-temperature vulcanizing rubber (RTV) that is used to bond the tiles to the shuttle airframe high enough to make the tiles fall off.

However, the RTV is underlain by a substantial aluminum structure, which can absorb a lot of heat. This huge heat sink allows the RTV, as well as the aluminum structure, to stay well below their design limits of about 200°F.

If you take a block of shuttle tile material and place it on top of a very good insulator and begin heating the front face, the back face will in only a few seconds heat up to a temperature equal to that of the front face. This is because the tile material has a very low resistance to temperature rise (in other words, a low specific heat), so that what little heat does reach the back face raises it to a high temperature. On the space shuttle, the aluminum structure is able to absorb a lot of heat before its temperature will go up (aluminum has a high specific heat and thus makes an excellent heat sink). Tom provided a more down-to-Earth example. When you put food in a cooler, you also add ice because the ice, as it melts, absorbs approximately 150 Btu per pound (of ice). The insulation in the cooler acts to slow the rate at which heat leaks in, so that it takes a long time for enough heat to leak in to melt all of the ice. The ice is the heat sink; heat that is absorbed by the ice is heat that is not raising the temperature of the food. Without the ice, or other heat sink like the food, the interior temperature will match the exterior temperature in a relatively short period of time because the heat capacity of the air and the inner liner is relatively limited. In the same way, a space vehicle's heat shield must slow the flow of heat into the interior of the vehicle, but flow it will. The heat that does make it through must be absorbed by something that will not reach high temperatures or allow other parts of the interior to reach high temperatures.

Going back to the early NASA programs, the Mercury capsule used a heat-sink heat shield, but lighter ablative heat shields had been the choice for the Gemini and Apollo programs. Ablators work by vaporizing and eroding, so that the material that erodes away carries with it some of the heat load. So in a sense, the ablator material can't "stand the heat," but it can keep the spacecraft cool. That is actually how the wooden heat shield on the Chinese Fanhui Shi Weixing (FSH) reentry vehicles worked. Wood can't withstand directly the temperatures of reentry, but for that relatively short time it can resist those temperatures by gradually eroding.

Ablators make use of the engineering principal of counterflow. In many industrial processes, it is desirable to heat, or dry, or wash a large quantity of material, with the minimum expenditure of heat, drying air, or washing water. The principal is simple. For example, if a material needs to be dried, it is moved in one direction along a conveyor, with drying air flowing in the opposite direction, so that the almost-dry material meets the driest air, and the wettest material meets air that is almost saturated with water vapor.

Another example, as Tom explained, would be a large pile of very dirty clothes that need to be washed very thoroughly with the minimum amount of water. The solution would be to set up a series of washers, where the dirty clothes would be taken from one washer to the next, in sequence. First, the dirty clothes would be placed in a washer with the dirtiest water, then they would move to a washer with slightly cleaner water, and so on, until finally they would go into a washer with fresh water. Meanwhile, the fresh water, after it had washed the cleanest load, would be moved downstream to the next cleanest washer, and so on until finally the last washer would have its dirtiest water replaced with water only slightly less dirty. This method would use the least amount of water by getting the maximum cleaning utility from all of the water in the system, from clean to very dirty. In a reentry heat shield, this principal can be used to minimize the flow of heat from the exterior into the structure by moving a cooling substance towards the hot oncoming atmospheric gases (actually plasma, as Tom pointed out).

The Chinese wooden heat shield worked that way, too. As the wood heated, a carbon ceramic char formed on the outer surface, and the volatiles, or fluids, in the wood behind the char flowed up through cracks in the char. Heat was radiated away from the charred surface, and the interior was kept cool by the outward movement of the cooler heat-absorbing volatiles flowing towards the hot side.

How much heat it takes to simply evaporate a pound of wood or a pound of the bathroom caulk-like material used in the Apollo heat shield can be easily determined. This can then be compared to the amount of heat required to evaporate and erode the same material when it is used as an ablator—in other words, when the flow of vapor is used to carry away the heat. For a typical ablative material, it requires three to five times more energy to erode a pound of the stuff as would be expected by merely considering the latent heat of vaporization. Again, this is caused by the counterflow principal at work. For this reason, ablators were the favored solution for the early reentry vehicles. They provided a lightweight solution to the reentry heating problem, and if the ablator were a material that also formed a ceramic surface—such as a carbon char—which would also reradiate additional heat, then it could be even lighter.

For ballistic missile warheads, which are designed to maintain their high velocities all of the way to impact or detonation and thus see very high temperatures, carbon-carbon materials—with a heat of ablation of about 5000 Btu per pound—are the preferred solution. The heat shield of the Galileo probe that entered Jupiter's atmosphere had used a carbon-carbon heat shield that was only a few inches thick.

ALL of these solutions were, however, intended for expendable vehicles. In 1960s, a lot of work had been done on reusable reentry heat shields employing high-temperature metals such as columbium. However, the metal heat shields had several major problems. They invariably had high weights per square foot, and their ability to withstand high temperatures was limited. They also were prone to erode in an oxidizing atmosphere, which they would encounter during reentry. This led to a search for some sort of reusable ablator, because ablators yielded the lightest-weight, simplest heat shields.

A reusable, or replaceable, ablator would be something that could be applied to the reentry heat-shield surface of a vehicle, for example, by spraying it on. Then, after landing, the charred ablator would be stripped off and new material applied before the next flight. This approach was tried on the X-15 on its speed-record flight, using MA-25S developed by the Martin Company. Unfortunately, the ablator proved to be almost impossible to remove from the vehicle after it had been charred by reentry heating. This led to an almost instant consensus that replaceable ablators were not feasible. No further efforts were made in this direction, nor any attempts to develop a heat shield that employed the ablator principal in a reusable fashion.

The abandonment of ablative cooling meant that the choices for high-temperature reentry protection were limited to ceramic tiles or some type of metal shield. Metal shields have the problems already mentioned: weight and limited maximum temperature capability. The space shuttle designers thus concentrated almost exclusively on ceramics, which have higher-temperature capabilities than metal, are lighter, and do not—in theory—require refurbishment between flights.

As amazing as the shuttle tiles are in many ways, they have turned out to be a major disappointment. Their sensitivity to erosion should the shuttle fly through a rainstorm—or even worse, through hail—remains a serious operational limitation. Metal tiles do at least have the advantage of being able to fly through a rain cloud without damage. The relatively brittle ceramic tiles also made it difficult to attach them to the airframe by any mechanism other than adhesive bonding. And, as Tom had been told by one aerospace manager, "we don't do bonding," or at least they don't like to, because of the difficulty of ensuring the integrity of the bond.

As amazing as the shuttle tiles are in many ways, they have turned out to be a major disappointment.

All of this contributed to the tiles' biggest drawback—the 30,000 hour required for reconditioning between flights. Each tile is checked to be sure that the RTV adhesive that bonds it to the aluminum substrate is still intact. Second, it is necessary to waterproof the tiles so that they don't become soaked with rain while the vehicle is waiting on the pad. In addition, there are numerous holes and doors in the tiles, which need to be inspected and repaired. The shuttle tiles are reusable, all right, but stripping them off and replacing them with some disposable system after every flight probably would cost less and take less time between flights. The shuttle tile material was usually considered to be the logical starting point for a reusable launch vehicle. Improved tile materials were available and—for a relatively simple ballistic reentry—could be used, but the basic problems remain.

More promising for Tom's upper stage, because of its lower heat loading, was a new, three-dimensional woven blanket material developed at NASA Ames Research Center. This material had been designed for use on the shuttle to replace tiles in areas where the aerodynamic loads were low enough that a completely rigid surface was not required. It consists of a base layer, a top layer, and a web layer that is woven back and forth between the top and the base to form a structure with triangular openings; all three layers are woven simultaneously. The outer layer could be made of silicon carbide, whereas the intermediate layers could be made of other, more flexible high-temperature fibers such as Nextel 440. The interstices would be filled with quartz wool. It appeared that it would be possible to weave a one-piece hemispherical heat shield for the base of the hydrogen tank, which could be fastened to the sides of the vehicle with Nextel straps, woven or sewn into the bottom of the sheet, as well as around the central engine opening at the bottom of the tank. A second, smaller, deployable heat shield could then be used over the engine opening.

They were also examining a simpler, related concept, borrowed from the design of high-temperature ceramic blankets for furnaces. Battens, constructed of quartz felt, could be formed into blankets and then folded into pleats. Metal fasteners would hold the bottoms of the pleats to the underlying heat-shield structure.

Although blankets seemed to be a practical approach, Tom was not entirely happy with it, and was searching for a solution that would keep the temperatures in the structure even lower, allowing the use of materials that would be easier to work with in the fabrication of the heat shield. In addition, for the engine compartment it was going to be difficult to place a heat sink behind the heat shield. Without a heat sink, as Tom had explained, the interior side of the heat shield would rise to the temperature of the outside surface rather quickly. That would result in a lot of heat radiating into the engine compartment. Although Tom thought this could be dealt with by relatively straightforward methods such as spraying water on the engines and interior of the compartment, he was actively investigating a system that appeared to be more robust.

One possible approach, which had been suggested by Boeing during the Heavy Lift Launch Study in the early 1970s, was to simply cool the inside of the heat shield by spraying it with water and using the evaporation of the water to absorb the reentry heating. For a typical ballistic vehicle with a total reentry heating load of approximately 2500 Btu per square foot, about 2.5 pounds of water per square foot (the heat of vaporization of water being 1000 Btu per pound) would be needed to cool the heat shield. In addition to the several pounds of water, a metal structure and a plumbing system for supplying the water to the heat shield would be needed, increasing the total weight. Even though that would amount to rather a lot of weight, the water-cooled heat shield appeared to be quite viable because the vehicles considered in this study were very large. As you increase the size of a launch vehicle, the weight grows as the cube, but the area of the vehicle, and therefore the heat shield too, only grows as the square. A vehicle that weighed 10 million pounds—100 times the weight of the AM&M vehicle—would have a heat shield that was only 20 times larger in area. The area and the weight of the heat shield on the Boeing vehicle therefore were effectively only 1/5 that of the heat shield for the proposed vehicle. Of course, the greater mass of the vehicle behind each square foot of the heat shield—the so-called ballistic coefficient or the number of pounds (more properly slugs) per square foot—meant that the Boeing vehicle experienced a higher heat load. Nevertheless, the smaller relative size of the heat shield kept the weight down to manageable levels. And, such a system would be very reusable. So, as a backup, Tom had some of his engineers designing a water-cooled heat shield. However, current estimates were that it would weigh approximately 1500–2000 pounds. Tom's design goal for the heat shield was 600 pounds, and 900–1400 pounds of payload was not going to be given up without a fight.

Tom was more interested in a concept called transpiration cooling, which has been used in some rocket-engine combustion chambers. In the 1950s, the Germans had proposed using a transpiration-cooled heat shield for reentry, which was based on the idea of using water essentially as an ablator. Transpiration-cooled structures could handle much higher heat loads than any known material, and so had subsequently been considered for various hypersonic leading-edge applications, which are notoriously difficult to cool. Transpiration cooling, as the name implies, is like sweating—the surface is made porous with many small, closely spaced holes, which allow a cooling fluid, typically water, but sometimes gaseous hydrogen, to flow into the boundary layer that overlies the surface. Because this

Transpiration-cooled structures could handle much higher heat loads than any known material...

fluid flow carries heat away from the vehicle, transpiration cooling is not unlike ablative cooling. However, no structure is used up, only the cooling fluid. So while the Boeing heavy-lift design required 1 pound of water to absorb 1000 Btu, a well-designed transpiration system might require only 1/3 or 1/5 of a pound of water for every 1000 Btu absorbed.

The challenge with transpiration cooling, though, was to design a lightweight structure that would contain all of the fluid flow channels required to deliver the fluid over the entire surface of the heat shield. But a more difficult problem was ensuring a uniform and adequate distribution of the cooling fluid. Any disruption or shortfall in the cooling flow could result in a catastrophic burn through of the heat shield. Although the aerospace industry had almost 50 years of experience with reentry systems, it was still more art than science to predict exactly what the heating profile on a new heat shield would be. This meant that significant safety margins would be required, at least until flight tests had demonstrated exactly what the heat loading would be.

TO overcome these difficulties, Tom had been brainstorming with his small team of reentry specialists. They were led by a relatively young man, Pat Nichols, who had worked on various "black" programs, as well as spending a short stint with the shuttle processing contractor. He was widely read and had a master's degree in thermodynamics from Cal Tech. Pat and his group had come up with several innovative approaches that could be tested relatively cheaply, and, if they appeared to hold promise, scaled up and flight tested on the prototype. The first one used the basic transpiration concept. However, to mitigate the potential problem of not having enough water flow where the heating was most intense, they had developed a concept using a face sheet of a shape memory metal such as Nitinol, a nickel-titanium alloy. Thousands of pores per square foot were formed in the sheet. These were then closed by plastic deformation. The shape memory metal was designed to have a transition temperature of between 150 and 350°F, so that when the surface of the heat shield reached that temperature in any particular spot the pores would begin to open, allowing more water to flow. In this way the heat shield was self-adjusting—wherever more heat was applied, more pores would open to allow more coolant to flow to the surface.

Another approach involved using a chemical compound with a relatively high weight of water bound in hydration. On the ground, or when exposed to the vacuum of space, the water would remain chemically bound, but when heated during reentry the water would be released and so provide a transpiration or ablative cooling mechanism. In this way, a very lightweight heat shield might be constructed.

This is the basic concept used for fire protection in homes and office buildings. Fire walls are constructed of ordinary drywall, or plasterboard, composed of gypsum (also known as plaster of Paris) with paper sheathing. Plaster of Paris, as every school-age child or artist knows, can be made by mixing calcium sulfide with water to form a paste that rapidly hardens into a relatively solid, dense, chalk-like material. The interesting thing is that the process is reversible—upon heating, the water is released, and you are left with dry calcium sulfate powder. This release of water, if the drywall is exposed to fire, results in transpiration cooling of the wall, delaying the progress of the fire through the wall. A 1/2- or

3/4-inch-thick drywall barrier can stop a fire for half an hour or more, depending on the intensity of the fire.

It seemed to Pat and his team that this concept could be used to build a reentry heat shield. Of course, covering the bottom of the vehicle with plaster of Paris would be too heavy, not to mention being prone to flaking away. But the principal was there. If the appropriate material could be found that held its water of hydration tightly, so that it would not evaporate in the vacuum of space, but *would* give it up during reentry, and, if the material could be readily rehydrated by spraying it with water, or even by simply letting the vehicle stand in a humid atmosphere, then they would have a lightweight heat shield that could be recharged and reused. Pat planned to run some initial tests using readily available materials like magnesium sulfate, gypsum, or zeolites, which can contain up to 40% water in their molecular structure. If the basic concept worked, Pat was sure that they could turn it over to the chemists to find one or more compounds that had a higher percentage of water by weight, perhaps a lithium or boron hydrate.

Tom also wanted Pat and his group to look at using hydrogen for cooling instead of water. Hydrogen, well appreciated by rocket-engine designers for its wonderful cooling capabilities, has a specific heat of 3.4 Btu per pound per degree Fahrenheit. This means that 1 pound of hydrogen can absorb almost three-and-a-half times as much heat per degree of temperature rise as water and almost 15 times that of aluminum. With the vehicle's expected total reentry heat pulse of about 2500 Btu per square foot, and by allowing the hydrogen to be heated to 1000°F, less than 1 pound of hydrogen per square foot of heat shield would be required. A very lightweight magnesium heat shield could be then be built, allowing hydrogen to absorb the heat at 1000°F as it flowed through channels in the heat shield. However, the downside of hydrogen cooling was the large additional volume of liquid hydrogen that would be required—perhaps a 100 cubic feet—which would have to be stored while the vehicle was in orbit. Another problem was the possibility of the vented hydrogen igniting as the vehicle entered the lower atmosphere.

Alternatively, hydrogen could be used for transpiration cooling, either through a shape memory metal shield as with the water system, or by thermal release from a chemical storage medium, such as had been developed for use in hydrogen-powered cars. In theory, a lightweight compound that stored hydrogen could be recharged in much the same way as the hydration scheme, but at substantially lower weight.

Another novel technique that they were exploring was use of a heat-absorbing phase-change material. Phase-change salts are well known for use in thermal storage systems. The original space station design had called for using a phase-change material to store solar energy for use during passage through Earth's shadow. However, this system was ultimately dropped, for reasons of perceived technical risk, in favor of the more conventional photovoltaic solar panels. For the heat shield, Pat's team was considering lithium hydride, which can absorb 1100 Btu per pound, although at a relatively high transition temperature of 1270°F. The lithium hydride could be packaged between two metal foils and would absorb heat at 1270°F. Or, instead of metal, the outer layer could be replaced by a thin ceramic layer, which would reradiate perhaps the majority of the incoming heat, with the remainder being absorbed by the lithium hydride.

Finally, they might end up combining several of these approaches, such as by using a thin layer of ceramic flame-sprayed onto the surface of a beryllium heat shield, underlain by a phase-change layer that would cover the aluminum tank wall. Tom's favored approach, though, remained the Ames three-dimensional woven blanket, in combination with a rechargeable hydrate compound, which would allow the blanket material to operate at lower temperatures. Most of this effort now was focused on a tailored zeolite to hold the water.

Tom was more than willing to try whatever ideas his team could come up with, especially those that could be tested at relatively low cost. Breadboard heat shields (i.e., simple prototype samples) can be tested somewhat crudely with an oxygen-acetylene or hydrogen-oxygen torch, but many of the reusable ablative/transpiration schemes relied on the low atmospheric pressures at which the heat is applied during reentry, in order to provide the maximum cooling capability. During reentry, maximum heating takes place well above 99% of the sensible atmosphere. Therefore, an arcjet facility was required to thoroughly test most of the concepts. Arcjets use an electric arc to produce a high-temperature, low-pressure plasma that is accelerated and impinged onto a heat-shield model. They can thus simulate relatively accurately various reentry conditions. The availability of arcjet facilities at any one of several government laboratories made this possible and affordable. The inclusion of a flying prototype in the program also meant that more than one concept could be flight tested, providing a real proof of the pudding.

The inclusion of a flying prototype in the program also meant that more than one concept could be flight tested, providing a real proof of the pudding.

Although Tom planned to select a final heat-shield approach before going into production, he also wanted to keep the vehicle design flexible enough so that more than one heat-shield concept could be used on production vehicles, in case there developed any long-term maintenance or reliability issues. With the initial system, that might not be obvious for the first year or so of operations. That was definitely going to be a part of the overall vehicle design philosophy. As with any aircraft, some systems would sooner or later be redesigned based on actual flight experience and retrofitted into the existing fleet. Tom was constantly reminding the engineers to do their best to provide for the possibility of retrofits of the heat shield, the reaction control system, the parawings, the engines—everything—in fact, except the basic airframe.

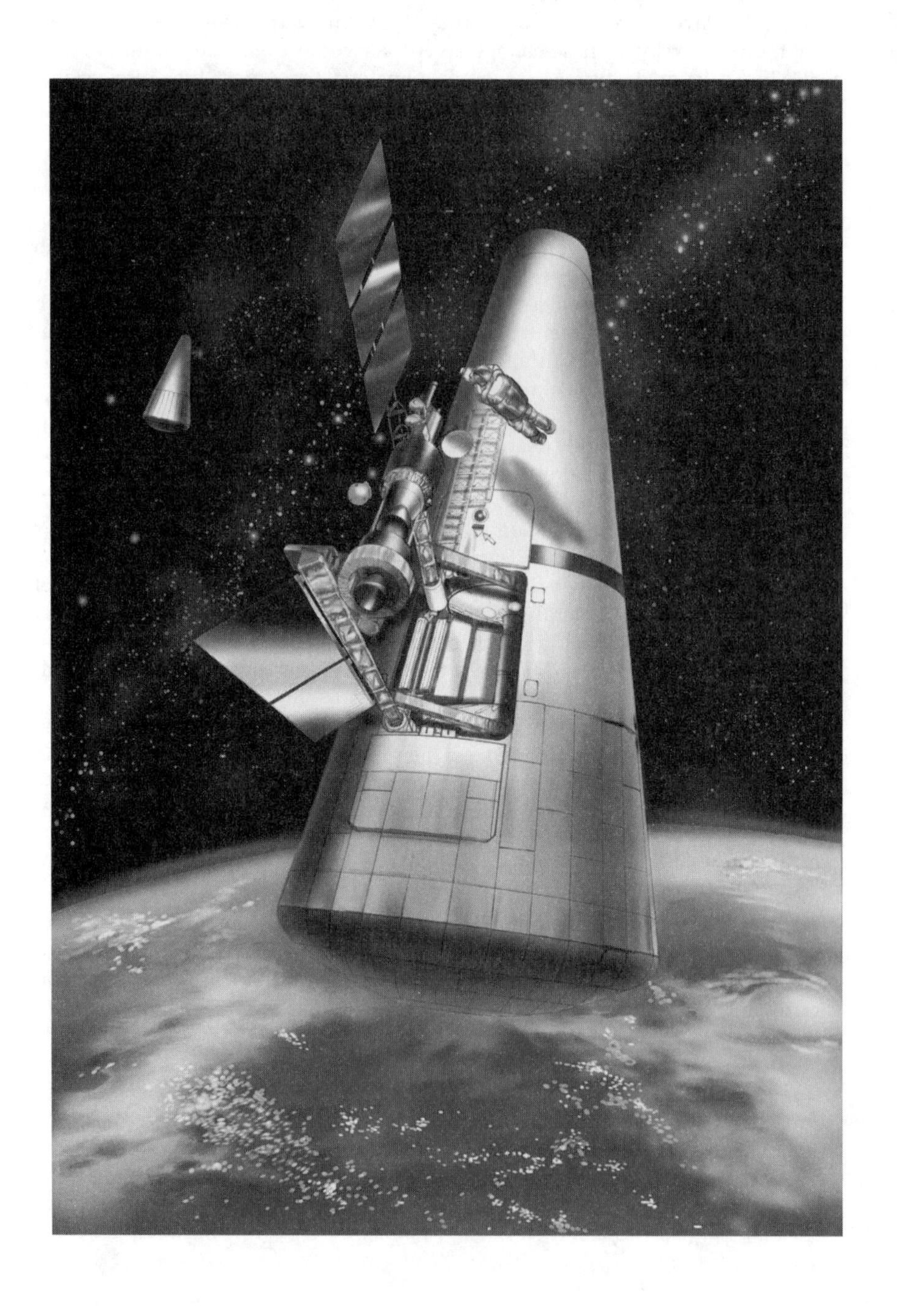

CHAPTER 11

Balloon Tanks, Fracture Mechanics, and Friction Stir Welding

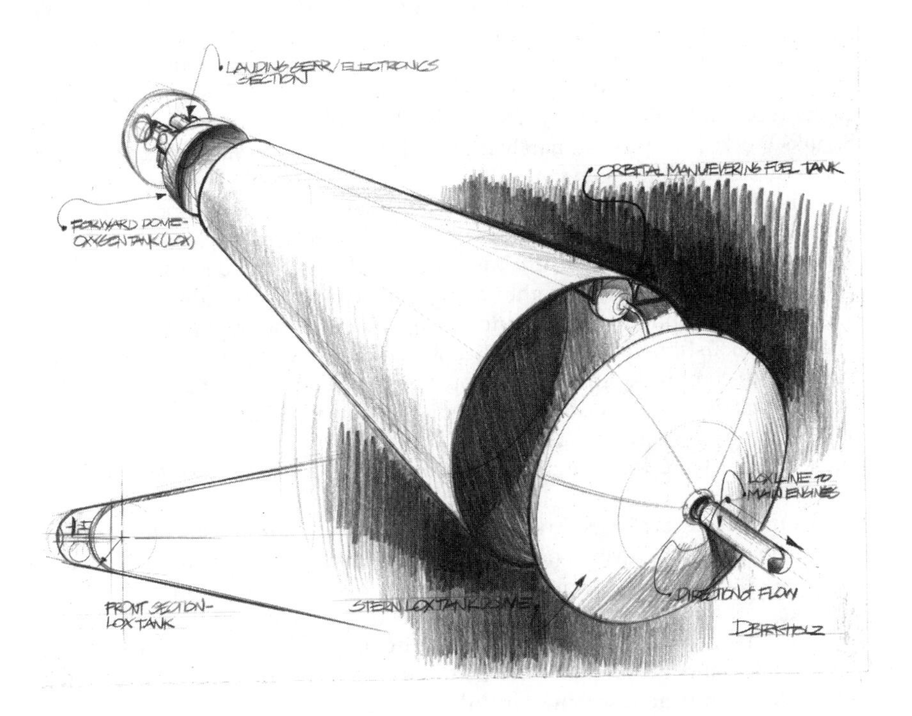

THE discourse on thermal protection systems was a bit more than I had bargained for, but it certainly left me with a better appreciation of the problem—and the level-headedness of Tom's approach to solving it. I was still curious, though, as to why composites didn't even seem to have been considered. After all, virtually all of the NASA and contractor studies of the past 10 or 15 years took for granted that composites, typically some advanced types, were absolutely necessary for a next-generation reusable launch vehicle. Bracing for another "fire hose" of information, I went ahead and asked Tom why he had passed them over. He agreed that the use of composite tanks has been widely studied and promoted in the past 20 years, especially after the problem of making a LOX-compatible composite tank had been solved to almost everyone's satisfaction. However, composite structures suffer from a number of drawbacks that never receive nearly as much press as their advantages. First of all, composites do not have the thermal conductivity or the homogeneous properties necessary for a good heat sink that can absorb the heat loads from reentry. From our discussion of thermal protection systems, I now knew just how important that was.

The second problem with composites is the difficulty of fabricating large pressure vessels. Burt Rutan had demonstrated that it was possible to lay up large structures by hand, relatively cost effectively, but for pressure tanks the numerous joints required by Rutan's technique make for a lot of uncertainty in the design and production process. The high-strength cases used so successfully for solid rocket boosters require extremely expensive tooling and very large autoclaves, and solid rocket cases are relatively simple structures compared to a reusable cryogenic stage. All of the structural details required for joining the various plumbing fixtures, access ways, and other features needed in a reusable cryogenic stage would make the design and fabrication of a composite structure significantly more difficult—and expensive. The death of the NASA/Lockheed Martin Skunks Works X-33 was ultimately triggered by the well-publicized failure of their composite liquid-hydrogen tank.

Finally, the plain truth is that, although they often show impressive weight advantages over metallic structures at the beginning of an aircraft or launch vehicle project, by the time the design is well into the details they usually end up with little or no weight savings. All of those little details of designing a workable, buildable structure eat up most of the supposed advantages. The Beech Starship was perhaps the most famous, or infamous, example. It was the first all-composite business turboprop aircraft to make it into production, but the Starship was an almost immediate commercial failure. Weight growth in its composite structure resulted in such a limited range that only about 50 of the planes were built, and most of these were scrapped less than 15 years later.

Commercial airplanes have indeed been using more composite structural parts over the past 25 years or so, particularly in the wings, tails, and interiors. However, they continue to rely on aluminum for the primary structures in the fuselage and the wings. Overall, composites are just too technically challenging for large structures. They are not necessarily any better in terms of weight or performance and are perhaps decidedly worse when maintenance, long-term fatigue life, and heat-absorbing capabilities are considered.

Thus Tom decided early on to stick with metal, and aluminum was his metal of choice. He went on to explain the two most widely used methods for building rocket propellant tanks. The first employs "isogrid" aluminum plate. This is fabricated by starting with a thick aluminum plate, sometimes as much as an inch and a half thick, and milling away more than 90% of the metal to get a plate with a thin skin plus an integral grid of upstanding reinforcing ribs, typically in a triangular pattern. This pattern provides the lightest configuration because it employs the triangular truss form, long recognized for its structural efficiency. The term isogrid refers to the fact that the plate acts much like an isotropic material—in other words, with uniform properties in all directions.

Isogrid plate allows the designer freedom to increase the stiffness of the aluminum plate where necessary so that the tank will be able to withstand concentrated punch loads. It also allows for increased thickness at the edges to compensate for the decrease in material strength caused by welding. Typical alloys used for tank fabrication are the high-strength aerospace alloys such as 2219. In the past decade, aluminum lithium alloys—with higher strength but a density 10% lower than those of other aluminum alloys—have come into use. Isogrid construction was first used by McDonnell Douglas on the Thor IRBM, and later in the Delta family of launch vehicles. Lockheed Martin also used it in the space shuttle external tank.

Another approach, developed by General Dynamics/Convair, and also used by the British on their Blue Streak launch vehicle in the 1960s, involved fabrication of the tanks from very thin cold-rolled, high-strength stainless steel. These so-called balloon tanks, with walls as thin as 0.030 inches, made possible the low mass fractions of the Atlas booster and Centaur upper stage. Cold-rolling stainless steel produces a very high strength sheet, but, as with the 2000-series aluminum alloys, welding weakens it. The welds in the Atlas that joined the stainless steel sheets to form the cylindrical hoops needed to be twice as strong as the welds that connected the cylindrical hoops together. Using stainless steel, which spot welds quite nicely, meant that the hoop welds could be easily reinforced by spot welding stainless-steel doublers over them on the inside of the tanks. The resulting tanks were strong and exceedingly light. But, as the name implies, balloon tanks rely on internal pressure to prevent buckling and to make the tanks rigid. The Atlas tanks were stabilized for ground handling by maintaining an internal pressure of 3–5 psi.

These so-called balloon tanks, with walls as thin as 0.030 inches, made possible the low mass fractions of the Atlas booster and Centaur upper stage.

During flight, all liquid-propellant launch vehicles depend on internal pressure in the tanks—typically 20–50 psi—to support thrust and launch loads, as well as to feed the propellants to the engines. Balloon tank design carried this to the extreme and used pressure stabilization to support ground and handling loads as well. Because they thereby eliminated separate stiffeners and numerous other structural elements, the Atlas tanks are among the simplest and lowest-weight tanks ever constructed. In spite of this, the technique was frowned on by many in the industry, particularly by the German rocket engineers who had been brought to the United States after the Second World War as part of Operation Paperclip and had been responsible for the design of early U.S. Army missiles and the NASA Saturn launch vehicles. They never felt comfortable with balloon tanks. Tom told of reading about an amusing anecdote from the early days of the Atlas program. Charlie Bossart, the chief engineer responsible for the Atlas, used to keep a 7-pound fiberglass mallet handy and whenever possible would try to coax a visiting member of the NASA German contingent to try their hand at damaging a test tank by hitting it with the mallet. Even more amusing than the Germans' amazement at their inability to cause any damage was watching them duck as the mallet flew wildly from their hands after they struck the tank.

Forty years of successful Atlas flights failed to convert the doubters. In the late 1990s, when the Atlas was being redesigned to use the Russian RD-180 engine, Lockheed Martin (who had absorbed Convair earlier) dropped the balloon tanks in favor of heavier, less robust, isogrid aluminum tanks.

TOM did not have such prejudices. He was looking for a way to build a lightweight tank that would be good for hundreds—and he hoped thousands—of cycles, and he knew that had never been done before. So he approached the problem with an open mind, and after careful consideration had opted to go with balloon tanks for their structural simplicity and light weight.

The first step had been to develop a weight budget for the vehicle, which set the target for the total tank weights. The orbital stage GLOW was 99,000 pounds—12,000 pounds of engines, structures and various vehicle subsystems; 5000 pounds of payload; and 82,000 pounds of LOX and liquid-hydrogen. With a 6:1 mixture ratio and an allowance of 5% ullage volume, the liquid-hydrogen tank would require a volume of 2775 cubic feet, and the liquid-oxygen tank 1400 cubic feet. A parametric analysis of various tank designs, with emphasis on the Centaur stage and the shuttle external tank, led to a conclusion that tank weights of approximately 3% of the weight of the propellants, or about 2500 pounds for the orbital stage, were within the state of the art. The total stage structure weight budget was 6600 pounds, including the thrust structure and payload bay. For his reusable vehicle, Tom took 1.5 times the minimum estimated weight for an expendable vehicle tank and set a minimum skin gauge of approximately 1/10 of an inch, which at about 1.4 pounds per square foot, came out to some 4400 pounds for both tanks. Minimum gauge was important both for resistance to fatigue cracking and to provide sufficient heat-sink capability for reentry heating. With 1.4 pounds per square foot of aluminum, an allowable 200°F temperature rise, and a specific heat of 0.215, the resulting heat-sink capacity would be, at a minimum, over 50 Btu/per square foot of tank surface area. The upper portion of the oxygen tank was going to be somewhat thinner because of the importance of keeping the center of gravity of the vehicle well down toward the base during reentry. The relatively low stress levels in the narrow portion of the cone would allow somewhat thinner gauges.

The next step was selection of the tank material. Typical aerospace design practice is to use high-strength alloys to build tanks, which are then pressure proof-tested and, if they do not fail, flown once. Tom, on the other hand, needed reusable tanks that could withstand many pressure and cryogenic cycles. This meant that strength was not his chief criterion, but rather, resistance to stress or corrosion cracking. So he turned to an aluminum alloy that is commonly used for saltwater boats and for cryogenic storage tanks, one that is both ductile and resistant to corrosion.

This meant that strength was not his chief criterion, but rather, resistance to stress or corrosion cracking.

As soon as it appeared that the Group of Seven would be giving the go-ahead for the project, Tom placed an order for a special heat of 100,000 pounds of aluminum alloy 5083 H116, in plates 1/8-inch thick. This alloy has good strength, good fatigue strength, excellent corrosion resistance, and good weldability. But Tom wanted the lightest tanks he could get, which meant that they had to use no more material than absolutely necessary, which meant that they needed material with the best and most consistent properties. Somewhere along the line, Tom had worked with a metallurgist who had shown him that the typical properties of an alloy could be improved by up to 10% and with better consistency if the purity of the starting material were tightly controlled. Therefore, Tom had specified electronics-grade ultrapure aluminum as the base material. He then worked with the mill to set up a special line where each plate, which typically has a thickness variation of plus or minus 10%, would be chem-milled to a standard thickness of 1/10 of an inch, plus or minus only 1%. By closely controlling the temper, they were also able to increase the elongation from the nominal 10% to something over 15%, as well as improve the yield strength to 32,000 psi and the ultimate strength to 45,000 psi. This all drove the cost of the material from $2.50 to $25.00 a pound, but considering that there would be less than 6000 pounds of aluminum in the orbital stage, it would only add somewhere in the neighborhood of 1/10 of 1% to the cost of the vehicle. Tom knew that if a material will stretch 10–20% before failing, most manufacturing flaws will result in a one-time plastic deformation, or stretching, of the material, which will redistribute the loads and stresses, rather than cracking and—in the case of a pressure vessel—rupturing. The trick for a good, long-life tank was to be sure that the metal did not yield every time it was cycled.

AT this point, Tom took me down to the materials section, where he introduced me to Thor Thingveld, his chief metallurgist, for a more detailed explanation of the materials issues involved in designing a highly stressed aluminum pressure vessel. Thor began by explaining that, in a thin plate that is under plane stress—like the surface of a balloon being stretched in all directions—the material is being stretched only along the surface or plane of the material. This stretching, however, causes the surfaces to compress the material in the interior of the plate. Therefore, depending on the thickness, the center of the plate could be under triaxial stress, that is, stress from all sides. This is important because triaxial stress leads to reduced ductility of a material, which can lead to crack formation and propagation. All manufactured structures contain, or at some point in their lifetime develop, small flaws or cracks. The study of how these cracks grow until the structure fails is called fracture mechanics. It is a particularly important topic when you are dealing with things whose sudden failure could be catastrophic, such as aircraft, submarines, nuclear reactors, and reusable launch vehicles. Strength is only one factor to be considered in the design of safe and successful structures; fracture toughness is critical for reusable structures, especially highly loaded reusable structures because it determines resistance to crack propagation and ultimate failure.

Most people aren't aware of it, but glass has a tensile strength that exceeds that of some steels. But, as everyone who has ever had their windshield hit by a rock knows, once you get even a small crack started in glass, it takes very little stress for the crack to propagate across the entire pane. The reason glass does not make a very good structural material is because it fails in brittle fracture, meaning that it does not stretch very much—hardly at all—before it breaks. And there are many ways in which a very small area of an object can be subject to very high local stress concentrations—a small hole, a crack, or even a scratch. If you take a pressure vessel in which the walls are in plane strain and drill a small hole through it, the material next to the hole will experience stress levels twice those in the material in the rest of the wall, far away from the hole. A thin crack has stress concentrations at the ends of the crack that are nearly infinite. This is essentially what happens in glass.

In a ductile metal, however, the material can stretch, allowing material that is distant from the stress concentration to take up some of the load. This explains why in most situations you can drill a hole through a steel plate or other ductile material without substantial loss in strength. The surrounding material takes the load that would have been supported by the material that is removed in making the hole. In most materials—including aluminum—a crack might or might not grow, depending on the stress in the material and the size and shape of the crack.

Typically, as a structure is subjected to many load cycles (in which loads are applied, then removed, then reapplied, over and over again), any extant cracks will grow. This can be aggravated by corrosive agents in the environment, which eagerly attack the more chemically active highly stressed material at the root of the cracks. Even in aluminum, depending on the particular alloy and the stress to which it is subjected, once a crack of a certain size forms, it can rapidly—even instantaneously—grow, leading quickly to explosive rupture in the case of a pressure vessel. This has been the main difficulty in the design of very lightweight aluminum tanks. Fracture mechanics is used to predict crack growth in terms of crack size and the number of pressure cycles experienced by the tank, and the simple solution (for expendable launch vehicles, at least) is to carefully inspect a tank after construction to be sure that it contains no cracks over a certain size. Then the tank is subjected to a pressure load that is higher than it will ever see during flight; if it does not rupture, then the cracks that are present but undetected must be smaller than the critical size, and the tank can be certified for a certain number of cycles.

Tom's criterion for a reusable launch-vehicle tank was that it should leak before a crack grew to the critical size where catastrophic failure was imminent. Thus, the choice of 5083 H116, an alloy with high ductility (characterized by an elongation of 15–20%) that will stretch a lot before failure, combined with excellent corrosion resistance. Fortunately, the tendency of aluminum to rip along a preexisting crack or tear is dependent not only on the ductility of the alloy and the actual stress levels within the material, but also on the thickness of the material. Very thin aluminum—such as aluminum foil—tears easily, and thick plates (over an inch or more) are prone to brittle fracture. It so happens that aluminum is toughest when the plate thickness is in the neighborhood of 1/10 to 1/4-inch thick—just the right range for the orbital-stage propellant tanks.

. . . aluminum is toughest when the plate thickness is in the neighborhood of 1/10 to 1/4-inch thick–just the right range for the orbital-stage propellant tanks.

In addition to the balloon tank concept and the selection of 5083 H116 aluminum alloy made from ultrapure aluminum stock, Thor had some other tricks up his sleeve for getting a lightweight, long-life tank. Because AM&M was buying the aluminum plate in a large lot, Thor 's technicians were going to carefully test a large number of coupons, or samples, to determine the exact properties of the aluminum in that lot. Then, rather than using handbook numbers for material properties such as yield and ultimate strength, ductility, critical crack size, notch sensitivity, and even coefficient of thermal expansion, the actual engineering properties would be used for the design calculations. After all, one of the reasons for paying 10 times more for the aluminum was to get and use the better material properties!

Next, the fact that the tanks were balloon tanks was going to be fully exploited. With pressure stabilization, nearly all loads would be borne in tension, which is the most easily modeled and analyzed type of loading. The precise analysis would allow the designers to come up with a very lightweight structural design by minimizing safety factors, which might be more properly termed uncertainty factors. (More will be discussed on that later.) Also, even the intertank volume between the cone-shaped liquid-oxygen tank and the hydrogen tank was going to be a balloon tank. This would allow this part of the structure, in the same way as the tanks themselves, to carry nearly all loads in tension. The intertank, which would contain the payload and crew station, would be maintained at a pressure of 5 psi over atmospheric pressure through the first minute of flight, during which time the intertank would experience loads over four times greater than it saw on the launchpad. After that point, it would be reduced to sea-level pressure of 14.7 psi, which at high altitude was more than sufficient to keep the structure in tension. The intertank was stiff enough to support the liquid-oxygen tank on the pad, but the structure could be made much lighter by using internal pressure to support the launch loads.

Another technique that Thor intended to employ was friction stir-welding, where a special rotating tool is forced down into the joint line between two pieces to be welded, thus generating frictional heat. The heat softens and plasticizes the material, without melting it, and allows the tool to be moved along the joint between two panels, transferring the plasticized metal from the forward to the rearward edge of the tool as it goes. The resulting solid-phase bond between the two pieces forms the weld. Because a stir-weld does not reach melting temperature, it does not have the usual effect of lowering the material properties of the aluminum. At the same time, the process also avoids typical weld-induced defects such as cracks, inclusions, and porosity.

The aluminum plate out of which the tanks were to be fabricated would be 100% inspected with thermal flow techniques and then chem-milled so that the thickness was within 1% of the nominal 0.10 inches. Thor's plan was to form the tanks by a combination of rolling and spin-forming the various pieces, and then stir-welding them all together. They would then create a computerized thickness profile of the entire tank by means of an ultrasonic scan, followed by thermal diffusion inspection for cracks. A small company in Wisconsin called Stress Photonics had developed this technique, in which beams of infrared light are played over the material being inspected; any minute surface cracks that exist will impede the flow of heat through the material, thus distorting the expected diffusion pattern. The system is capable of detecting very small surface cracks. Any surface cracks that were found in the tanks would be repaired by stir-welding.

Finally, the tank would be selectively chem-milled to remove any unnecessary material and therefore weight. Thor was still working on several approaches to chem-milling the tank. Chem-milling is a process that utilizes a basic solution to etch aluminum in a very controlled manner, where conventional machining would be too difficult, or too coarse. A maskant, or "resist," is used to protect those areas where you don't want material to be removed by the basic solution.

At this point the likeliest approach involved using a robotic system to lay down a layer of resist where needed, followed by the chem-milling process that would selectively etch away the metal wherever there was no resist present. This would be repeated in one-half thousandth-inch layers until the interior of the tank had been sculpted exactly for minimum weight.

TO support these design and manufacturing techniques, the structures group was developing detailed finite element analysis models for the vehicle. Tom explained that although finite element analysis (FEA) was useful as a design tool for predicting stress levels, you had to be careful not to rely too heavily on the results, because they were based on a model of the structure, not the real thing. In his view, FEA was most useful for correlating and interpreting experimental data from loads testing on actual hardware.

After the computer models were generated, subscale tanks, which had the same wall thickness as the production tanks, were fabricated. These tanks incorporated all of the various details that would give rise to stress concentrations—man-way access hatches, fittings for the propellant and pressurization lines, raceways for data and communication lines, brackets, and reinforcements. The tests involved pressurizing the tanks to a level that created the same stress in the subscale tanks as would be seen in the full-size versions. For one-quarter-scale test articles, the internal pressure needs to be four times the pressure that would be used in the full-scale tanks; for 1/10 scale tanks, the pressure must be 10 times greater. Data from these subscale tests were obtained by using full-field photoelastic techniques, which had also been developed by Stress Photonics.

The models were finally tested to destruction by increasing the pressure until they ruptured. This allowed verification of the ultimate load and also highlighted design details that were causing excessive stress concentrations. Because the lower dome of the oxygen tank and the upper and lower dome of the hydrogen tank were not spherical, there were compressive loads in addition to the tensile loads, and some stiffening was necessary. This was accomplished with thicker tank wall sections in some areas and a series of ribs welded to the intertank side of the domes. The ribs also did double duty by serving as the mounts for the payload positioning and support rails. The design of the stiffening features proved to be a headache for the structural engineers at first, as they strove to keep the weight down. But extensive testing with the subscale models gave them confidence to proceed to the full-scale tanks.

Thor's other big concern was fatigue life. Aluminum is one of those materials that will eventually break if subjected to a sufficiently large number of load cycles. Fatigue is a major issue for aircraft that are intended to have a useful life of 30 to 40 years. In the case of the venerable DC-3, the designers lacked complete confidence in the newer aluminum alloy they were using, and so they kept the allowable stress levels well below those used in modern design practice. As a result, the DC-3 has a nearly infinite structural life, which explains why 150 DC-3s are still flying in the U.S. alone, more than 70 years after they rolled off the production line. Modern aircraft can't afford the luxury of such overdesign. They are carefully tested for fatigue failure by cycling the fuselage through numerous pressurization cycles, so that they can predict fatigue life with a high degree of confidence. Sometimes, service life of an airframe can be extended by means of a very thorough inspection for any signs of cracking and recertification of the airframe if none are found.

For the orbital stage, Thor started with some of the subscale tanks. They were filled with water and cycled. So far they were achieving several thousand cycles before any cracks developed. However, testing with the more complicated combination of thermal as well as pressure loads was required to provide complete assurance that the tanks would have the lifetime required for a reusable vehicle. To do this with relatively short cycle times, a fog of liquid nitrogen was used to chill the LOX tank, which would not be insulated in the flight vehicle. The liquid-hydrogen tank, which would actually be insulated, was tested without the insulation so that it too could be brought to operational temperature with liquid nitrogen. Thus, relatively small volumes of liquid nitrogen could be used to rapidly cool the tanks. After cooling, infrared heat lamps were employed to bring the tanks up to expected reentry temperatures, followed by mechanical application of the landing loads. Because the tanks were necessarily pressurized with gas during this operation, a special reinforced concrete structure with a blowaway roof had to be built. A complete cycle took just over two hours, which meant that 12 flights could be simulated in one day, and 1000 flights could be simulated in three or four months—if none of the test equipment broke down. Thor also used subscale test articles, primarily the 1/10- and 1/4-scale tanks, to verify the design of every penetration or attachment to any structural element within the vehicle. Subscale testing really helped to hold down test time and costs.

BY means of this exceedingly thorough testing, Tom was aiming for a high level of confidence in the design so that they could build a vehicle with safety factors that approached 1.0, which to most engineers is no safety factor. Thor clarified the concept of safety factor for me. He explained that the calculation of stress in any engineered device is necessarily imperfect. The equations are based on idealized models of a physical reality, and as we have been discussing, minute cracks, imperfections in the material's crystalline structure, and small deviations from the assumed material shape can each create unpredictably high stresses within the material. So, for most structures, a safety factor is employed to compensate for the uncertainty about material properties and uncertainty about applied loads. A safety factor of two, for example, means that the you take the theoretical maximum stress that the material can withstand, and halve it to arrive at the highest stress that you will allow in your design.

Tom told me the classic story of the young engineer being directed to design a fixture for use in an automobile assembly plant, only to be told after he had completed his design to use materials with five or 10 times the necessary strength. The young engineer finished the job feeling that the whole exercise had been silly—until one day the fixture was run into by a forklift; it suffered no damage because it was so massively overdesigned. Overdesign was no option for the orbital stage—every pound of additional structure meant a pound less payload, and if they weren't very careful, the entire payload could all too easily evaporate. Better computer modeling and analysis of

Overdesign was no option for the orbital stage–every pound of additional structure meant a pound less payload.

the structures and loads alone would not suffice, although Thor and John Patterson, chief of the structures group, and their respective teams did their best to get as close as they could.

The solution to building a robust yet lightweight vehicle, again, lay in using a very ductile material. As aluminum, and many other metals, stretch, they actually become stronger. With the selection of custom-made, high-purity 5083 H116 aluminum alloy, Thor was able to predict accurately the increase in strength that would result from a certain amount of stretching. Coupled with the extensive testing program, this permitted Tom to sign off on a design using little or no safety factor, at least in the traditional sense. The relatively low-strength 5083 had sufficient ductility to stretch and gain in strength if subjected to unexpectedly high stresses; the safety margin was contained in the plasticity of the tank material.

The flight vehicle structures would be tested until yielding started to take place. During yielding, any stress concentrations would yield first and so end up in compression when the loads were removed, leaving the entire structure free of stress concentrations that might cycle through the plastic limit every load cycle. It was plastic yielding that would lead to crack formation in a highly loaded structure. If an unexpected stress developed—the proverbial forklift hitting the vehicle—the stretchability of the material would result in a one-time plastic deformation, providing an increase in strength where needed, albeit at some loss of ductility.

To support routine flight operations for the production vehicles, Thor was designing a detailed schedule of inspections and periodic overpressurization tests that would ensure the integrity of the tank for a given number flights. Undoubtedly, experience over time would allow them to increase the number of flights between major inspection and test points.

Thor was working on a nondestructive inspection technique using the principle that an electromagnetic wave or signal can be used to induce an ultrasonic signal in an object made of iron, steel, or other magnetic material. This technique was used to generate ultrasound for sonar early in the 20th century. In the 1990s, a technique that would allow the inspection of oil and gas pipelines and industrial steam pipes without requiring a physical coupling between the sensor and the pipe was developed. By placing a coil around the circumference of a pipe, including several magnets to amplify the effect, an ultrasonic wave was produced that could travel as far as 100 feet along the pipe. The same coil could then be used to detect the ultrasonic waves that would be reflected back by any defects in the pipe, such as very small areas of corrosion.

Of course, the AM&M vehicle's tanks were made of aluminum, and this technique would not work with nonmagnetic material. However, by plating a thin layer of nickel over a portion of the tank interior, extending completely around the circumference, the entire tank could be ultrasonically tested with a relatively simple coil and inducing magnet setup placed around the exterior of the tank. Less than 5 pounds of nickel would be required on the inside of the tank, and it could be deposited in much the same way as is routinely done in the manufacture of computer hard drive platters. With this simple test, the entire tank could be thoroughly checked for any pitting or cracks. Thor's idea was to perform this test after, or perhaps before, every flight.

CHAPTER 12
Enthusiasm Bubbles, Ejections, and Expander Cycles

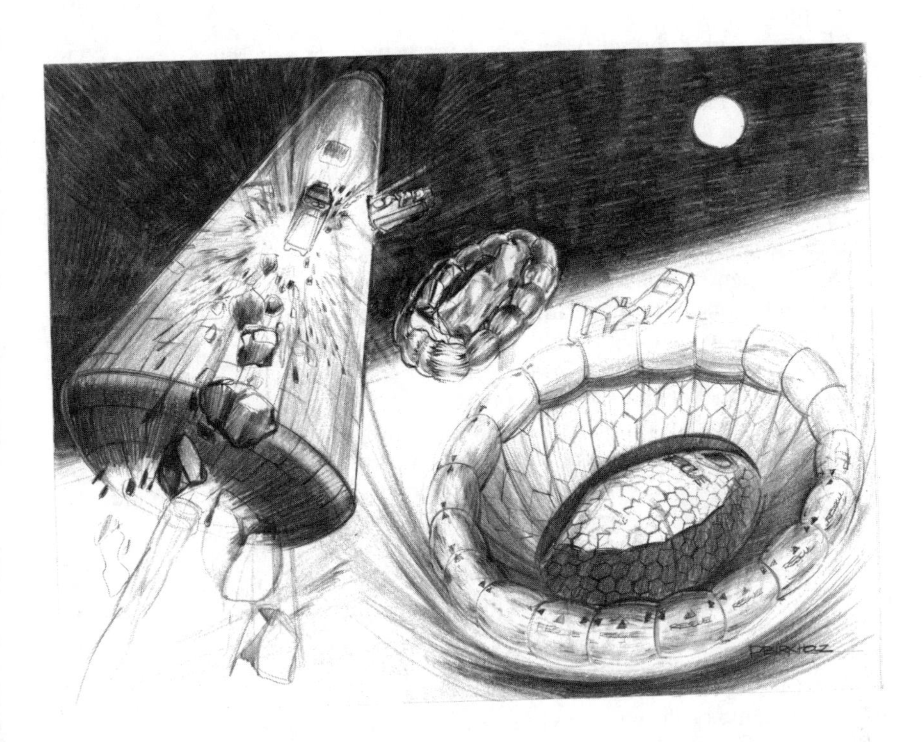

FOR some time now, John Forsyth had been working with Dick Stefan, an attorney and veteran of the commercial aircraft certification process. Together they had been making preliminary contacts with the Federal Aviation Administration (FAA) Office of the Associate Administrator for Commercial Space Transportation, or AST as it was known in office-symbol parlance. They had gotten to know the principal members of the staff, particularly the political appointees, but also the technical people who were assigned the job of drafting regulations and reviewing launch license requests. Prior to the major funding commitment, Forsyth had limited their activities to sizing up the personalities and gaining an understanding of the organization's mindset and the general direction they were taking in space transportation regulation.

AST had not yet been faced with the need to certify a manned, reusable orbital launch vehicle for commercial operations, but several of the X PRIZE contestants had submitted applications for manned suborbital flight to the X PRIZE altitude of 62 miles. Forsyth, through an intermediary, had actually provided $5 million to one of the X PRIZE contenders for the very purpose of exercising the system and testing the waters to get a feel for the regulatory environment. This experience enabled him to rate the regulatory uncertainty as relatively low when he put together the Group of Seven. As good fortune would have it, the test vehicle that they were now building had a first-stage trajectory that was remarkably similar to that of an X PRIZE vehicle and could take advantage of the regulatory regime set up to handle the various suborbital tourist vehicles under development.

Even though the predictions of a large market for suborbital tourist flights had not yet proved out, a precedent had been set for the operation of manned, rocket-powered launch vehicles to over 300,000 ft. So for the first-stage prototype, it was not going to be a major extrapolation beyond the suborbital vehicles in obtaining a launch license. The orbital second stage, however, would be breaking new ground for a private vehicle. Forsyth was hopeful that AST could relatively painlessly come up with some reasonable hybrid of the emerging suborbital regulations and the well-established regulations pertaining to conventional two-stage expendable launch vehicles.

When the major funding came through, Forsyth had brought Stefan onboard full time and had him begin right away the process of working in earnest with AST. Forsyth's intent was that, as the vehicle was designed, the regulators would feel like they too were part of the process and thereby be more likely to buy into the technical approaches that AM&M was taking. It was almost always the case that a regulatory agency for a new industry tended to become an advocate for the industry. It is, after all, a somewhat symbiotic relationship: without an industry, the agency doesn't have any reason to exist. Also, regulatory agencies are typically staffed by former members of the industry or related industries, and there is always the old "revolving-door" phenomenon, where they might expect at some point in the future to be employed in the industry they were regulating. Thus, by taking reasonable care to avoid any bruised bureaucratic egos, Forsyth didn't really expect any serious problems with regulation. To be sure, flying the prototype well before the production vehicle would provide the advantage that the regulations could be developed within the context of an experimental vehicle, but with an eye to eventual commercial production and operations, first in the United States and later in the rest of the world.

The first step was to overcome a certain amount of generalized skepticism within AST. The commercial launch-vehicle business had seen more than its share of flaky startups, as well as more than a few serious startups that had been underfinanced and never amounted to anything. The equally long list of NASA and military launch-vehicle programs that never got beyond the PowerPoint stage also tended to fuel this skepticism. After awhile, even the most sympathetic regulators had a tendency to pretty much figure that any new group of entrepreneurs that came in the door was equally unlikely to get anywhere. In spite of all that, Forsyth wasn't yet ready to make public the investors' very respectable financial commitment, as a tool to gaining credibility. AM&M's credibility was something he intended to manage very, very carefully.

The commercial launch-vehicle business had seen more than its share of flaky startups, as well as more than a few serious startups that had been underfinanced and never amounted to anything.

Part of his strategy was to do everything he could to make sure that every time a potential customer, government regulator, or supplier heard something about AM&M their estimation of the company and its vehicle would rise. This meant avoiding any overinflation of public expectations or impressions of the company. It was best, Forsyth felt, if at first everyone underestimated the seriousness and the viability of the venture. Then, as people followed AM&M's progress, they should tend to gradually gain more confidence in the company's ability to do what it said it was going to do and, he hoped, grow more interested.

Forsyth was a keen observer of the human condition and was well aware of the difficulty of changing someone's mind once they had formed a definite opinion about something. By purposely keeping expectations low at the beginning—and, more importantly, carefully managing their development—he hoped to prevent the formation at any point of widespread or generally accepted negative views of the company.

FORSYTH wanted people to start with an attitude of curiosity about AM&M, and then turn that into real interest in what the company was doing. Next, he wanted them to become intrigued, and then mildly surprised as they started to show real progress, proceeding to restrained expectations about what was going to happen next. Finally, things would no doubt get to the point of wild enthusiasm, and perhaps overblown expectations, after the prototype first flew to orbit. Much better to have that stage develop around the time the company was going out to the markets for additional capital, rather than earlier. He hoped to avoid the "bubble and bust" phenomenon, which seemed to hit all new industries at some point, by controlling the growth of the bubble of enthusiasm. Forsyth also wanted to be sure that there was a lot of real substance underneath that would support and sustain the industry if enthusiasm got out of hand and the bubble burst.

Undoubtedly, some of the early customers for the vehicle would overpromote their prospects and raise a lot of money, only to find that positive cash flow was just a few years too far downstream to prevent bankruptcy. It would probably be a lot like the dot-com bubble—fueled in part by widespread expectations of a global broadband market, which did come to be, eventually; it just came too late for most of the players.

But as far as AST was concerned, Forsyth considered it much more important to come across as a serious and technically competent operation than it was to convince the regulators that they were well funded and likely to make a lot of money someday. In this way, they could quietly build a relationship with the regulators based on the technical issues and lay the groundwork for a system of regulations that would be beneficial for AM&M and the industry which the Group of Seven intended to create. With this approach, the stakes—at least to the regulators—would seem to be relatively low. When the stakes are perceived to be high, bureaucrats sometimes lose the courage to make definitive decisions when they need to be made and take up defensive positions behind impossibly high standards. Or, in order to spread or deflect blame for anything that might go wrong, they drag too many other agencies into the process that don't really have any need to be there, which only serves to slow things down and provides more opportunities for someone, somewhere, to throw a monkey wrench into the process. No, it would be better, as AST laid the groundwork for regulating commercial space transports, if AM&M's vehicle was seen as somewhat abstract. It would be—at first—merely an interesting concept that might have possibilities, but in any event would provide a useful exercise as the agency felt its way forward in regulating the industry, which was the real reason for its existence: commercial, manned spaceflight.

WHEN considering all of the regulatory challenges, Forsyth was particularly pleased that Tom's vehicle had some real pluses with respect to flight safety, both for people in the vehicle and those on the ground. AM&M's launch vehicle—now designated the DH-1—compared very favorably with NASA's most recent reusable launch-vehicle concept, which was currently in the preliminary design phase. Compared to the space shuttle—still the premier manned space transportation system in the world—well, there was no comparison. Funding for the new NASA vehicle had only reached the $2 billion level, which was hardly enough to do the wind-tunnel testing for the hypersonic booster stage. Forsyth and Tom were both quite confident that, after another $1 or $2 billion was spent, the whole concept would quietly wither away, like all of the other next-generation NASA programs. At least the current NASA design formed a good straw man with which to compare the safety features of the AM&M vehicle. In the DH-1, both stages would have ejection seats for the pilots and also for the crew and passengers in the orbital stage.

The pop-up booster—Tom had come up with this name for the first stage—would experience ejection environments that actually were less severe than those for the SR-71, which in the worst case involved ejection into a 2500 mile an hour breeze at 80,000 feet. The vehicle's vertical trajectory meant that it would reach maximum aerodynamic pressure at about 35,000 feet altitude, moving at slightly supersonic speed; the maximum dynamic pressure would be roughly 600 pounds per square foot. These conditions were approximately those encountered during ejection from a supersonic jet fighter. As the vehicle climbed out of the atmosphere, air pressure would of course decrease, and although dynamic pressure increases with the square of the velocity atmospheric pressure falls off logarithmically. So, aside from the need for a full pressure suit, bail-out conditions improved in most respects as altitude increased.

The escape system for both the first and second stages was based on the Russian K-36D-3.5A ejection seat, a lightweight version of the seat used in the MiG-29S fighter. The design of the K-36D-3.5A is modular, making it relatively easy to adapt for the AM&M's vehicle. The basic seat weighs 156 lb; a harness, oxygen supply, and survival kit would need to be added.

The escape system for both the first and second stages was based on the Russian K-36D-3.5A ejection seat, a lightweight version of the seat used in the MiG-29S fighter.

For the first stage, the pilot's seat was mounted on a hydraulic platform behind a hydraulically operated door. A 5000-psi hydraulic accumulator supplied pressure to open the escape door and move the pilot and seat through the door and out of the vehicle. Then the ejection-seat rockets would take over to get the pilot quickly away from the vehicle. Once he had separated from the seat, it would be necessary to stabilize the pilot, in order to prevent the high spin rates that can develop during a fall through the highly attenuated upper atmosphere. Here, aerodynamic forces are strong enough to induce a spin, but the atmosphere is not dense enough to allow the pilot to control his attitude with body movements, as conventional parachutists and skydivers can do. Two systems were used for stabilization. The first was a very small, three-axis cold-gas attitude control system (ACS)—essentially a smaller, simpler version of the attitude control systems that are used to orient and maneuver spacecraft. This system would permit the pilot to control his orientation during what might be a 5-minute coast upward into space, followed by a vertical descent to the launch site. The second stabilization system would be used during descent, after the aerodynamic forces had built up to the point where the cold-gas system could no longer control the pilot's motion. It consisted of a drogue parachute, deployed at 100,000 ft; which remained attached until it was released at 8000 ft, thereby triggering the deployment of a parawing. With this, the pilot could then make a pinpoint landing at the launch site. At deployment of the drogue chute, the pilot plus full pressure suit and oxygen system weighed between 250 and 300 lbs. To provide further backup, an emergency reserve parachute—altimeter actuated at 3000 ft if the pilot were unconscious or disabled—was also provided. This chute would be disarmed by the pilot at 5000 feet if everything was going well, but it could be manually deployed if problems developed with the parawing during the last part of the descent.

On the orbital stage, ejection would be initiated by firing a linear-shaped charge to cut a hole in the side of the vehicle, followed by the ejection seat being shot out through the opening. The ejection and recovery sequence for the orbital stage pilot would be identical to those of the first stage for any bailout prior to staging. Ejection after that, however, meant that the pilot would have to survive a more or less severe reentry into the atmosphere. The good news was that the actual bail-out conditions were now pretty benign, the pilot being ejected into the near vacuum of space and therefore experiencing no significant aerodynamic effects. Therefore, upon ignition of the orbital stage engines the ejection sequence was modified so that the seat would bring along a 75-pound reentry package and also employ a lower speed, less violent ejection from the spacecraft. The reentry package would be pneumatically loaded into a hollow in the back of the ejection seat, allowing it to be automatically attached to the pilot's harness.

After ejection, the seat would separate from the pilot, and the same cold-gas system used with atmospheric ejection would provide control of the pilot's orientation, except that now it would put him into a slow, safe spin to minimize solar heating. When the pilot initiated an ejection, the vehicle's onboard navigation computer would download the bail-out conditions—in terms of position and velocity vector—to the reentry package computer. This information would then be used to time the deployment of the package.

The reentry system used ballistic modulation, which had been conceived in the 1950s and actually used by the Russians in the late 1990s for the recovery of payloads from space. The AM&M system consisted of a large Nextel 440 fabric disk, which was deployed by a coiled stainless-steel spring sewn into the rim. The low ballistic coefficient of the combined disk structure and pilot kept the heating load within the capabilities of the fabric. A hemispherical sponge-like cocoon would be deployed between the pilot—in his full pressure suit—and the disk.

The sponge, soaked with a water-alcohol mixture that could absorb the reentry heating load by evaporation, would actually keep the pilot at a cozy 70°F. The disk and pilot would initially be oriented by the cold-gas ACS, which was mounted on the pilot's chest. Once aerodynamic forces began to act, the pilot would be pushed forward relative to the edge of the disk, which served to give the system aerodynamic stability as a result of the change in shape—it now would look something like a badminton shuttlecock and thus have a lower ballistic cross section. Plunging into the atmosphere, the structure would experience the maximum heating rate at a point where it was experiencing relatively light aerodynamic loads. When maximum dynamic pressure, which corresponds to maximum *g* loads, was reached, the weight of the pilot would deform the shape of the device even more, further lowering its ballistic cross section and decreasing drag, allowing it to penetrate more readily through the atmosphere. As it became less blunt, this lower drag would reduce the total dynamic load, and spread it out, so that the maximum *g* load experienced by the pilot would be reduced from 9 or 10 to slightly over 6 *g*. Once he was below 100,000 feet, a drogue parachute would deploy, pulling the pilot away from the disk. The same parawing used in the first-stage system would likewise allow the pilot to make a controlled landing.

Thus, both stages of the vehicles had survivable abort modes during all phases of flight. Abort at any stage of flight was definitely a design goal of every manned launch vehicle since Vostok, but the required hardware had proven to be too heavy for the shuttle, and such abort systems tended to drop out even in more recent NASA concepts as a result of cost and weight impacts. Eventually, for on-orbit rescue from a disabled vehicle or of a disabled crewperson or passenger, Forsyth anticipated that an orbital rescue service would be available that could reach them within the 48-hour on-orbit capability of the DH-1 vehicle. At some time in the future, no doubt, a vehicle held in ready at a launch site on the equator could rendezvous typically within 2–4 hour. Such a mission might make a good job for the U.S. Coast Guard.

At some time in the future, no doubt, a vehicle held in ready at a launch site on the equator could rendezvous typically within 2–4 hours.

Both stages were designed with explosively actuated tank drains, so that over most of the trajectory the propellants could be dumped and the stages recovered. At this point, doing this for the oxygen tank was still somewhat problematic, as the explosive charge could conceivably ignite the aluminum tank walls and destroy the vehicle. But again, this was something that could be tested on a small scale. The engineers had hopes that by coating the inside of the tank opposite the point where the explosive charges were installed, ignition of the tank could be avoided.

The first stage employed five main engines in addition to four vernier engines and, because of its high thrust to weight, had one-engine-out capability over the entire trajectory and two-engine-out capability over most of the trajectory. Because of its high vertical velocity, the orbital stage, which would employ either six RL10s or, it now appeared likely, two RL60 engines, could operate a single engine to propellant exhaustion, and then reenter from any point in its trajectory. The vehicle would not have quite the flexibility of an airplane, which can often lose an engine and still continue to its destination, albeit at somewhat slower speed; the DH-1 would fail to achieve orbit if engine shutdown happened early in first-stage or second-stage flight.

Of course, premature engine shutdown is usually a euphemism for explosive disassembly, when applied to rocket vehicles. But, here too the DH-1 had some important safety advantages: the rocket engines of both stages in the production vehicle would employ the "expander cycle" developed by Pratt & Whitney for the RL10.

THERE are basically three types of turbopump-fed rocket engines. That is, there are three engine cycles used to provide the power to run the propellant pumps. The purpose of the pumps in a rocket engine is to force the propellants into the combustion chamber, where they combine and burn to form high-pressure gas, which by flowing out of the chamber and through the rocket nozzle produces the thrust that propels the rocket into orbit, or wherever it is bound.

Rocket-engine turbopumps are highly stressed pieces of machinery, handling incredible amounts of power per pound—something like 100 hp/lb for a space shuttle main engine. To do this job, they are aided by having available a large flow of high-energy propellants from which to extract energy and to serve as a working fluid. The oldest cycle, which was used in the engines of the Saturn V, is the gas-generator cycle. This involves taking some of the fuel, perhaps 2 or 3%, and burning it with some oxidizer in a separate combustion chamber (the gas generator) to produce a relatively cool, fuel-rich gas stream, which drives the turbine (or turbines), which in turn drives the pumps that feed the propellants into the chamber. After driving the turbines, the gas is exhausted overboard. This cycle is relatively straightforward in that the gas flow is relatively independent of the other engine system components, that is, the pumps, the injector, and the engine chamber. Variations on this cycle include employing a completely independent gas generator and burning propellants other than those used in the main engine. The V-2 used monopropellant hydrogen peroxide, catalyzing it to produce high-pressure steam to drive the turbines.

There are several problems with the gas-generator cycle, however. First of all, as the engine chamber pressure is increased to improve engine performance more propellants have to be used in the gas generator to produce the additional energy necessary to pump the propellants at higher pressure into the chamber. The exhaust from the gas generator produces very little thrust when dumped overboard. Thus, as you seek to improve the efficiency of the engine by increasing the chamber pressure, the overall efficiency at some point begins to fall as more of the propellants are required for the gas generator. As a result, the maximum chamber pressure that is reasonable for this cycle tends to end up in the 1500-psi range.

For gas generators that use the main engine propellants, there are potentially more serious problems. If something goes wrong somewhere in the engine and the fuel supply to the gas-generator decreases, or the oxidizer supply increases, the gas generator temperature will increase. This then causes the turbine to speed up, which (usually) makes the problem worse by increasing the flow of oxidizer, further increasing the gas temperature, until the turbines are destroyed either by overheating or overspeeding.

A failure mode might start out as a relatively minor, noncatastrophic problem, such as a burn-through of a few of the cooling passages in the engine chamber wall. Initially, this would result in excess fuel being dumped into the chamber at the site of the burn-through; the excess fuel would then cool this area, preventing further damage. However, in a gas-generator cycle engine, the resulting lower pressure on the fuel side of the engine can cause the gas-generator to go off its mixture ratio, leading eventually to rapid disassembly of the engine.

A second engine cycle, employed on most Russian engines from early on, and used in the space shuttle main engines, is the staged combustion cycle. In the U.S. version, all of the fuel and some of the oxidizer are run through a preburner, whose combustion products are used to drive the turbines and then dumped into the engine thrust chamber and burned with the remainder of the oxidizer. In the Russian variation, they burn all of the oxidizer with some of the fuel in the preburner and dump the oxidizer-rich combustion products into the engine chamber after they drive the turbines. Staged combustion engines have the advantage that—because all of the propellants eventually end up in the engine—you can extract as much energy as you need to run the pumps without affecting engine performance. This means you can build engines with much higher chamber pressures, which yields higher performance and higher *Isp*s. This is especially important for booster engines that operate at sea level, where the expansion of the exhaust gases in the engine nozzle is limited by atmospheric backpressure.

Here again, though, there are problems. Staged combustion engines, which at first seemed to be the future of high-performance rocket engines, lack a certain amount of engineering robustness. There are many points within the shuttle SSMEs (space shuttle main engines) that have caused concern over the years. For example, to allow the SSME to achieve its extraordinarily high operating chamber pressure of 3200 psi, the oxygen pressure must be raised to 7400 psi and the hydrogen to 6500 psi. To achieve these pressures with a staged combustion engine, the hydrogen must be preburned with some of the oxygen, run through the oxygen and hydrogen turbines, and then the hot, high-pressure turbine exhaust is dumped into the chamber. The high temperatures in the oxygen and hydrogen powerheads have caused turbine-blade cracking and excessive bearing wear. Also, the oxygen injection tubes are exposed to the high-velocity, high-temperature, and high-pressure turbine exhaust gases at the inlet of the injector. These engines are sensitive to overheating as well; an anomaly in the injection of either the liquid oxygen or hydrogen into the preburners can cause a hot spot that will overheat a turbine blade. The loss of a single turbine blade would lead rapidly to destruction of the entire engine, and with the extremely high chamber and turbine manifold pressures containment of the blast would not be possible.

Staged combustion engines, which at first seemed to be the future of high-performance rocket engines, lack a certain amount of engineering robustness.

On the other hand, no SSME has ever failed in flight. But all through the early 1980s, especially before the *Challenger* accident, most people with at least some real knowledge of the shuttle's inner workings held their breath during every launch—with exploding SSMEs uppermost in their minds. At first, the average life of an engine was only three or four flights. With the upgrade to the advanced Pratt & Whitney powerheads, other redesigns, and minor improvements following the *Columbia* accident, SSME reliability and durability had improved—somewhat. Still, these could be considered reusable engines only in the same way that the shuttle could be considered to be a reusable vehicle. With enough maintenance, overhauls, and rebuilds, they could be reused, but they weren't particularly cost effective. With a flight rate of one or two flights per year per vehicle, though, the engines were adequate.

Although Tom admired the engineering that had gone into the SSME, he had long been of the opinion that for a reusable launch vehicle, which for AM&M meant up to 300 flights a year, not three, a more robust engine was needed. For the DH-1, Tom turned to an engine cycle that had been developed early in the space program, in fact for the first successful liquid hydrogen-liquid oxygen engine, the RL10. It is called the "expander" cycle because power to drive the turbines is taken, prior to any combustion. Hydrogen is an amazing gas. Its low molecular weight allows it to absorb a lot of energy with relatively low change in temperature. An expander-cycle engine uses the liquid hydrogen fuel to cool the engine by flowing through a series of narrow tubes or grooves in the chamber and expansion bell (nozzle) walls. After absorbing heat from the walls, the heated and now-gaseous hydrogen is used to drive the turbines for the pumps. The hydrogen at the outlet of the engine cooling passages is typically at less than room temperature. (In fact, the hydrogen keeps the engine chamber cold enough that liquid water actually drips from the inside of an RL10 bell—while it's running!) Therefore, the turbomachinery operates at room temperature or below; the only high-temperature gases are in the combustion chamber, and the turbine working fluid does not present any possibility of overheating or overspeeding the turbine. Thus, the expander cycle is inherently a relatively low-stress rocket engine, with high reliability and excellent reusability.

In the RL10, with a chamber pressure of approximately 600 psi, the hydrogen turbopump raises the hydrogen to approximately 1700 psi. The high-pressure hydrogen flows through the cooling passages, through the turbines, and then into the engine injector, where the now-high velocity hydrogen gas is very effective at breaking up and thoroughly mixing with the incoming liquid oxygen stream, yielding a combustion efficiency of over 99%. Liquid oxygen, at 900 psi, is supplied by a turbopump driven by a gear train coupled to the hydrogen turbine. Because of the liquid oxygen's lower pressure and higher density, the volumetric flow of oxygen is only slightly more than one-third the flow of hydrogen, so that the power required for the oxygen turbopump is only one-sixth the power required for the hydrogen pump (lower pressure × lower volume) and can be efficiently transmitted by a gear train. The expander cycle also has the advantage that it can handle a wide range of operating conditions. The RL10 engines have been modified to operate from 10–100% of design thrust levels, which allows for a lot of flexibility in flight and vehicle applications.

An off-the-shelf RL10A3-3A, with 16,500 pounds of thrust, has been run for over 30 minutes without any maintenance. Pratt & Whitney is confident that one of their RL10s could make 25 flights before some maintenance might be necessary on the bearings and the gear train and perhaps 100 flights before the gear train would need to be rebuilt. The ultimate life of the current engine design was estimated to be about 130 flights, based on cryogenic thermal cycling of the bell.

After extensive discussions with Pratt & Whitney, Tom was confident that the basic engine could easily be redesigned for even greater reusability by using a thrust chamber with milled slots rather than brazed tubes for the cooling passages and by using a separate turbine to drive the oxygen pump directly. A gear train would be used only to keep the hydrogen and oxygen pumps linked together for mixture ratio control. Such an engine should be capable of 100 flights between overhauls, with a total life expectancy of around 500 flights.

Happily, in addition to its durability, the expander cycle is also inherently safer, especially in the RL10 configuration, where the oxygen and hydrogen pumps are geared together. This coupling together means that it is essentially impossible for the chamber to ever become oxidizer rich, which could cause destruction of the chamber through oxidation or more likely by overheating. There is also a direct link between the integrity of the engine chamber and the power supplied to the pumps. If any cooling passages should leak, the hydrogen flowing into the chamber would cause additional cooling around the site of the leak. This would in turn lower the hydrogen pressure to the turbine, reducing the flow of both propellants and causing the engine to operate at lower pressures, thus checking any tendency for the engine to "run away." The first and only in-flight failure of an RL10 in the late 1990s—after over 30 years of flight history—was due to lack of braze coverage for the chamber structural support jacket. Updating the design to use milled channels instead of tubes for the fuel passages allows the structural jacket to be integrated directly into the chamber.

Of course, with any engineered device, there is always a downside. For expander-cycle engines, as you increase the thrust at some point there is just not enough heat available from the chamber walls to be absorbed by the hydrogen so that you have enough power to drive the turbines. As the engine is scaled up, the available heat from the thrust chamber only increases as the two-thirds power of the thrust. Pratt & Whitney has been able to extend the thrust range for the expander cycle up to 100,000 pounds by using a heat exchanger that heats the liquid hydrogen with the output of the turbine, so that the hydrogen is prewarmed before entering the cooling passages. Even engines with 200,000 pounds of thrust appear to be feasible. The latest Pratt & Whitney engine using this cycle—the RL60—was now being considered for the current DH-1. The expander cycle could also be made to work with other fuels, and Pratt & Whitney was currently working on a 100,000 pound thrust LOX-methane engine for use on the first stage.

FROM a regulatory perspective, then, with public safety the main concern, the DH-1 was about as good as it could get. The rocket engines would use the safest and most reliable cycle available; furthermore, the pop-up booster configuration and trajectory would provide continuous crew-escape capability during all flight phases, intact continued flight to orbit with one engine out on either stage, and intact recovery with all engines out over the majority of the flight path. And, the fact that the first stage always remained over the launch site meant that it would not present a hazard to populated areas downrange.

The only remaining area of uncertainty was overflight of populated areas by the upper stage, which depended on the probability that the upper stage might experience a failure that would result in an uncontrolled ground impact. Of course, the stage would weigh only 100,000 pounds fully fueled (far less than most airliners), and the fuel consisted mainly of hydrogen, which tends to dissipate rapidly in the event of a crash. All in all, this meant that the maximum damage that could be expected from an orbital stage impact was approximately that same as that from one of the midsize business jets that overfly populated areas by the thousands every day.

In the end, permission for overflight by DH-1s was going to depend on a statistical analysis of actual flight experience, which might require a lot of flights before the confidence level reached a point that was at least within an order of magnitude of that of a business jet or airliner.

However, the unique trajectory of the vehicle tended to reduce the risk to any particular point on the ground. If, for instance, the orbital-stage engines failed just after ignition, and if the propellants could not be dumped, it would take about 250 seconds for the vehicle to impact the ground. This means that as the orbital stage accelerates, the ground impact point moves much farther over the ground than the vehicle does. A single second of thrust, which would accelerate the vehicle to over 32 feet per second horizontal velocity—would result in moving the instantaneous impact point about 1.5 miles downrange during the 250 seconds before impact. For a typical airplane, flying much closer to the ground, the ground impact point moves about 0.15 miles during 1 second of powered flight. The DH-1 would have an instantaneous impact point that moves 1.5 miles per second after 1 second of powered flight. After 10 seconds of thrust, the impact point would be moving downrange at more than 15 miles per second.

The result is that, for purposes of determining the probability that a vehicle will crash into any particular point as it flies over, the orbital stage is effectively flying 10–100 times faster than an aircraft, and so exposes any particular ground facility to a much smaller risk because the exposure is for a much shorter period of time—tens to hundreds of times shorter.

For a typical launch vehicle, the danger is much greater from a crash near the launch site, which means launches must be conducted far from major population centers. The DH-1's trajectory, though, would result in the vehicle traveling so rapidly over even a nearby population center that it would be in no greater danger than a population center thousands of miles away from the launch site. Most of the ground track for a typical flight would be over unpopulated territory, and so the probability of a crash is less important than the risk posed by a vehicle as it passes over a population center. Dick Stefan believed that if the vehicle reliability could be shown to be at least 0.999—the minimum design goal for the DH-1—it would have an overflight risk level comparable to that of a business jet.

Dick Stefan hoped that he could work out in advance with AST a basic system reliability target that would provide a level of safety for overflight comparable to commercial aircraft operations, and once the statistical data for vehicle reliability were available, approval for launch over populated areas would be granted. Of course, the same statistical data would allow an assessment of the potential liability for insurance purposes. Preliminary negotiations with the underwriters were already showing that liability insurance costs of under $1000 per flight were possible—even over densely populated areas—once sufficient flight data were in hand.

So, as the design of the DH-1 progressed, the regulators were tutored and educated and included in the design process and the safety analyses. Forsyth and Stefan put a lot of effort into developing a regulatory framework that was forward looking, especially in terms of providing a logical and reasonable process for assessing vehicle safety and in laying the groundwork for a legal regime that would allow spacecraft to be launched from anywhere, with launch-site restrictions that were no more onerous—and perhaps even less so—than those for a major airport. They were confident that, once adopted in the United States, these rules would eventually become the worldwide standard. This was critical to ensuring that the market for vehicle sales would be large enough to provide the return on investment that Forsyth and the other members of the Group of Seven were anticipating. Even small, densely populated, landlocked countries would be able to gain access to space, and the prestige associated with that capability would, they hoped, spur vehicle sales.

Even small, densely populated, landlocked countries would be able to gain access to space, and the prestige associated with that capability would, they hoped, spur vehicle sales.

CHAPTER 13
Gasoline, Alcohol, Kerosene, or Liquid Methane

I had heard that the propellants for the prototype first stage were not going to be the same as for the production vehicle, so I stopped by Tom's office to ask him about it. He started to explain his choice of propellants for the prototype—98% hydrogen peroxide and liquid methane at a mixture ratio of 15 pounds of hydrogen peroxide to 1 pound of methane—but I immediately asked, "Why not liquid oxygen and methane? That is, I understand, the baseline for the production first stage, isn't it? And actually, I have been wondering why you didn't specify kerosene instead of methane—kerosene seems to be the hydrocarbon fuel of choice for everybody else."

Tom responded, "Those are good questions, and they deserve good answers. You know, I think you could use some historical context—the story of rocket propellant development is rather fascinating—and understanding it is important to what we are doing here. In fact, why don't we go talk to Peter down in the propellant chemistry section?"

Peter Nakamura was seated at his desk, poring intently over a technical journal, when Tom knocked on his office door. After introductions, Tom said, "Peter, our chronicler here needs some background in rocket propellant history, and you're the local expert." Peter waved us to his guest chairs, sat down in his own, and began. He started with a review of the fundamentals of rocket propulsion.

"The important thing for a rocket," Peter explained, "was the velocity of the gases that a particular propellant combination—consisting of a fuel and an oxidizer—could produce in a properly designed engine. Newton's third law says that for every action there is an equal and opposite reaction, and so the faster the gases leave the rocket engine, the more thrust it develops per pound of propellants burned. This exhaust, or jet, velocity is proportional to a parameter referred to as specific impulse, or *Isp* (pronounced eye-es-pea). The exhaust velocity divided by the acceleration due to gravity gives the *Isp*, in units of seconds."

I had some familiarity with these terms from earlier discussions of the rocket equation, which, as I had learned, was the key to the whole problem of designing a launch vehicle.

Peter went on to explain that there are basically two things that determine how fast a rocket can go: the velocity of the exhaust jet and the ratio between the starting weight and the burnout weight. Thus, there are two characteristics of rocket propellants that are critically important to the performance of the vehicle: the *Isp* or jet velocity they can produce, and the propellants' densities, which determines how much you can get into the propellant tanks. Other properties need to be taken into account as well, depending on the application: cost, toxicity, suitability for engine cooling, and ease of handling. But *Isp* and density drive the performance.

Other properties need to be taken into account as well, depending on the application: cost, toxicity, suitability for engine cooling, and ease of handling. But *Isp* and density drive the performance.

Jet velocity is proportional to the square root of the temperature of combustion as divided by the molecular weight of the exhaust gases. So a good rocket-propellant combination will have a high energy of combustion for a high-temperature flame and combustion products with low molecular weight. If you double the flame temperature, you increase the jet velocity by a factor of the square root of two, for about a 40% increase. If you halve the molecular weight, you also increase performance by the same factor.

LONG before anyone had ever built, fired, or flown a liquid-propellant rocket engine, the Russian space pioneer Konstantin Tsiolkovskii had determined that the most energetic combination of propellants that was practical for use in a rocket engine was liquid oxygen and liquid hydrogen. This combination has a high energy of combustion, 6900 Btu/lb, and the combustion product is water vapor, with a nice, low molecular weight of 18. However, difficulties in obtaining, storing, and using the −423°F liquid hydrogen were such that the early rocket pioneers avoided it. They did use liquid oxygen—which has a higher boiling point, higher density, and relatively low cost, and so is much easier to work with than liquid hydrogen—combined with a hydrocarbon fuel.

In the United States, Robert Goddard, who built and flew the first liquid rockets, chose gasoline, whereas the Germans settled on alcohol. The German combination had the advantage that you could add water to the alcohol, which, while reducing the flame temperature, also reduced the molecular weight, so that the rocket engine could be run at a lower temperature with a relatively small loss in performance. This lower temperature proved necessary, given the engine chamber cooling technology (film dump) that was used on the V-2. Goddard also found that liquid oxygen and gasoline burned at too high a temperature and, with less success, worked on ways to inject water into the chamber to reduce the temperature.

Peter lent me an entertaining and educational book, *Ignition* by John Clark, that describes the history of the search for the best military propellant. For military purposes, the goal was a combination of propellants that could be stored at room temperature—or, preferably, over a wider range of temperatures experienced under field conditions—and which was also hypergolic. The term hypergolic is of German origin, and it means that when the propellants are combined they ignite spontaneously. This of course simplifies the design and improves the reliability of missiles by eliminating the need for an igniter. It turns out that nitric acid makes a pretty good oxidizer, especially the red-fuming kind (RFNA), containing some dissolved nitrogen tetroxide, which is responsible for the reddish-brown fumes. It has, however, a ready tendency to corrode almost anything it touches. Fortunately, it was discovered that by dissolving a small amount of hydrofluoric acid in it you could reduce its corrosiveness dramatically. It is hypergolic with kerosene, but better performance can be achieved by burning it with hydrazine (N^2H^4). However, hydrazine freezes at a temperature too high for military purposes. Eventually, a combination called aerozine-50 was found—consisting of 50% UDMH (unsymmetrical dimethyl hydrazine) and 50% hydrazine—that has a low-enough freezing point, high density, and good storability. This fuel mixture, with nitrogen tetroxide as the oxidizer, was used in the Titan missile and by some of the early space launch boosters. The Russians, however, typically used straight UDMH and nitrogen tetroxide in their missiles.

The first missiles built after WWII used liquid oxygen and kerosene, which was quickly established as the propellant combination of choice for missiles like Redstone, Titan I, and the Atlas. This combination gave good performance, and, even though it included cryogenic liquid oxygen, both propellants were readily available. However, the problem with kerosene is that, like gasoline, it is a mixture of various hydrocarbons. Knowing exactly what that mixture is isn't terribly important when kerosene is used in lanterns or heaters. But the variations of two important properties—density and coking temperature—do present problems for rocket engines. They both depend on the batch of kerosene you happened to get. Density is a problem because when you fill up the fuel tank you don't necessarily get the weight of propellant that you expected, and how much propellant (by weight) you have onboard has a lot to do with the performance your rocket will achieve.

Coking is what happens when the temperature of the kerosene reaches a certain value; it will then produce a tarry residue that forms deposits on whatever it happens to be touching. This is a problem because the preferred method of cooling the engine chamber for most rockets is to run the fuel through cooling channels or tubes that make up the walls of the thrust chamber. So, if kerosene is the fuel and coolant, and the temperature gets too high, carbon deposits will form on the hot walls of the cooling tubes. These deposits reduce the thermal conductivity of the wall, which can lead to burn-through of the inner wall of the rocket chamber. Another problem with commercial kerosene is the sulfur content, which can result in corrosion problems.

To solve these problems, ordinary kerosene was replaced as a rocket fuel by a special grade with a specified high coking temperature and more consistent lot-to-lot density. It was cleverly named RP-1 (Rocket Propellant-1). On the downside, this meant that the fuel was no longer as readily available, or as cheap. However, by this time most large engines were designed to run on kerosene, and although many other hydrocarbon fuels have been considered and tested none have succeeded in replacing RP-1.

The persistent coking problem and the higher cost of RP-1 meant that Tom did not consider it suitable for a reusable engine in a reusable vehicle. He had, after thoroughly working it out with Peter and the engine design group, settled on liquid methane, also known as liquefied natural gas, for the first stage. Liquid methane is cheap, readily available, and easily purified so that it is extremely consistent in density from lot to lot. It has a very high coking temperature, making it excellent for cooling, and a low enough molecular weight makes for a very respectable *Isp*. They also briefly considered ammonia, which doesn't coke at all, is an even better working fluid, and costs about the same as methane. Although ammonia had been used on 199 flights of the manned X-15 with nary a problem, in the final analysis its toxicity made it an also-ran.

The persistent coking problem and the higher cost of RP-1 meant that Tom did not consider it suitable for a reusable engine in a reusable vehicle.

Methane and ammonia are also good working fluids for driving the turbines in an expander-cycle engine. However, in order to get enough heat into the methane for driving the pumps, the engine combustion chamber would have to be somewhat longer than the optimal length, and a heat exchanger would be used to recover heat from the methane turbine exhaust. Thus, as methane flowed from the propellant tank it would first be warmed in the heat exchanger and then used to cool the engine chamber walls. The heated methane, now at about 1200°F, would be expanded in one turbine to drive the methane pump and in another separate turbine to drive the oxygen pump. Both pumps would be geared together, not to transmit power, but to ensure that the mixture ratio between the propellants remained constant. The methane could thus be heated to a temperature sufficient to provide the energy for running the turbopumps with no danger of coking. Methane has a density of only about 40% that of kerosene, but given the higher mixture ratio (oxygen to methane) the overall average density of the propellants is not that much worse than for kerosene and liquid oxygen.

Peter explained that liquid oxygen and liquefied natural gas were chosen for the first stage primarily on the basis of cost: oxygen was only about $0.03 per pound and liquefied natural gas less than $0.20 per pound. This meant that the total cost of propellants for the first stage would only be about $10,000. Pratt & Whitney had been given a contract to develop a 120,000-lb-thrust expander-cycle engine for this stage. For the orbital stage, liquid methane might be useful for planetary missions, especially if you wanted to refuel on Mars, and it offered two other major advantages. It is six times denser than liquid hydrogen, and much easier to store for long periods in space. So Tom also had Pratt & Whitney working on a version of their RL60 engine that would run on liquid oxygen and either hydrogen or methane, although methane would provide a somewhat lower maximum thrust.

FOR the prototype first stage, however, Tom and John Forsyth had decided that they needed to fly it earlier, rather than later, and they couldn't wait for the new engine. This presented a problem. Peter explained that there was no suitable, reusable engine available in the thrust category they needed. So for the prototype, they had to come up with a reliable, affordable, short-term solution: a purpose-built pressure-fed engine. But which propellant combination to use for this engine wasn't obvious, as Peter explained. The factors that drove the selection for the production vehicle didn't necessarily pertain to the prototype program, and so the team surveyed the available propellants from this perspective.

In the rocket business, familiarity with particular propellant combinations tended to lead to consensus that only one or a few were worthy of consideration after some point because of the experience and also the equipment and infrastructure that had been developed for those particular propellants. The long life of the kerosene/RP-1 and liquid oxygen combination was the principal case in point. Another combination that had also developed a loyal following was hydrazine/monomethyl hydrazine and nitrogen tetroxide. This pair had the advantages of high density, low corrosiveness, and a high *Isp* at around 320 seconds. It had been widely used on expendable launch vehicles, unmanned spacecraft, and satellites, as well as for the space shuttle orbital maneuvering engines and reaction control system. Hydrazine can also be used as a monopropellant, meaning it doesn't need a separate oxidizer. Monopropellants are attractive because there is only one propellant to be concerned with, which greatly simplifies the design of the engine system. However, the high cost and time required to safely process the shuttle systems that used these propellants—not to mention heightened awareness and concern for toxicity and environmental issues—had led, by the turn of the century, to a renewed interest in less toxic propellants for reaction control systems and even for on-orbit maneuvering systems.

One such propellant was hydrogen peroxide, which had been used on the X-15 and Mercury capsule reaction control systems, but had then fallen into disfavor. Hydrogen peroxide (H_2O_2) has long been used onboard submarines and naval ships to power torpedoes and emergency generators and is relatively safe. It can be decomposed catalytically into high-temperature steam and gaseous oxygen and thus can serve as a monopropellant. When burned stoichiometrically with almost any hydrocarbon or other fuel, it provides even better performance, with an *Isp* only somewhat less than that of hydrazine and nitrogen tetroxide.

Hydrogen peroxide in the prototype boost stage had the other advantage that its density is high enough to allow meeting the tank weight targets required for a pressure-fed stage, and it can also be easily used to generate the pressurizing gas for pressurizing its own tank.

One problem with hydrogen peroxide, though, was that industry essentially stopped making the propellant-grade stuff around 1970. Oh, it was still a major industrial commodity, used for cleaning and sterilizing, particularly in food applications, because it left no harmful residual; it was also widely used for bleaching paper pulp. Millions of tons of the stuff were produced and consumed, in concentrations of 30, 50, and 70% hydrogen peroxide in aqueous dilutions, which cost approximately $1 a pound for the hydrogen-peroxide content. But for rocket-propellant use, 90% or higher concentrations are needed.

The problem was the method of manufacture. Back in the 1960s it was manufactured by electrolysis of an aqueous solution of sulfuric acid or acidic ammonium bisulfate to form peroxydisulfate, which was subsequently processed by hydrolysis. A lower-cost process had been developed, wherein a hydrocarbon carrier was treated with hydrogen and then successively oxidized to hydrogen peroxide and then stripped from the hydrocarbon. This nonelectrolytic process was considerably cheaper, but the downside was contamination of the end product with small amounts of hydrocarbon.

Combining a hydrocarbon with hydrogen peroxide produces a less stable mixture, meaning that it tends to spontaneously break down into water and oxygen. To overcome this problem, various stabilizing compounds can be added. Unfortunately for propulsion engineers, these stabilizing compounds contaminate the catalyst beds. They also make it difficult to reach the very high concentrations of hydrogen peroxide—98 to 99%—that are most desirable for propellant applications. Actually, these high concentrations are inherently quite stable without stabilizers, in the absence of hydrocarbon contaminants. However, within the last few years the renewed interest in hydrogen peroxide as a nontoxic rocket propellant meant that 98% hydrogen peroxide was again available, although at a high price of almost $10 per pound.

Even so, for the limited needs of the prototype flight-test program, that wasn't excessive. Tom had let a contract with TRW for a 108,000-pound thrust, ablatively cooled hydrogen-peroxide-propane engine, based on their famous lunar module descent engine, the LEMDE. TRW was well along with the development of this engine, with an estimated cost of $30 million for 15 engines, and a design life of 10 flights each.

Hydrogen peroxide could also be used as a monopropellant for the first-stage vernier landing engines, on the production vehicle as well as the prototype. It makes for easy throttling—all you have to do is open a valve until the desired thrust level is reached—and the catalyst bed provides superreliable ignition. Hydrogen peroxide could also be used to run a small turbine to power the hydraulic actuators for thrust vector control of the first-stage main engines and the landing engines. These factors, and the inherent safety of hydrogen peroxide, would help keep the development cost of the first stage low, without compromising reusability and reliability.

A 10,000-pound thrust hydrogen-peroxide catalyzed engine was already available, and flight testing was underway using just these as the vernier engines. The prototype had tankage to provide the engines with 10,500 pounds of hydrogen peroxide for ascent and 10,000 pounds for landing, making a total of 20,500 pounds of propellant. The lightweight test stage weighed 25,000 pounds empty and 45,000 pounds fully fueled, which gave it a delta-V capability of about 2500 feet per second. This was enough to boost the vehicle to an altitude of a few thousand feet, so that landing techniques could be developed and the flight control and landing systems debugged. Thus, a useful flight program was in progress only one year after go-ahead, without the need to wait for the main engines. It also provided the opportunity for AM&M to go through the process of having its first manned vehicle evaluated for a launch license with the benefits of the reliability and safety of an all-hydrogen-peroxide propulsion system.

Thus, a useful flight program was in progress only one year after go-ahead, without the need to wait for the main engines.

CHAPTER 14
Design Reviews, Prototypes, and Parawings

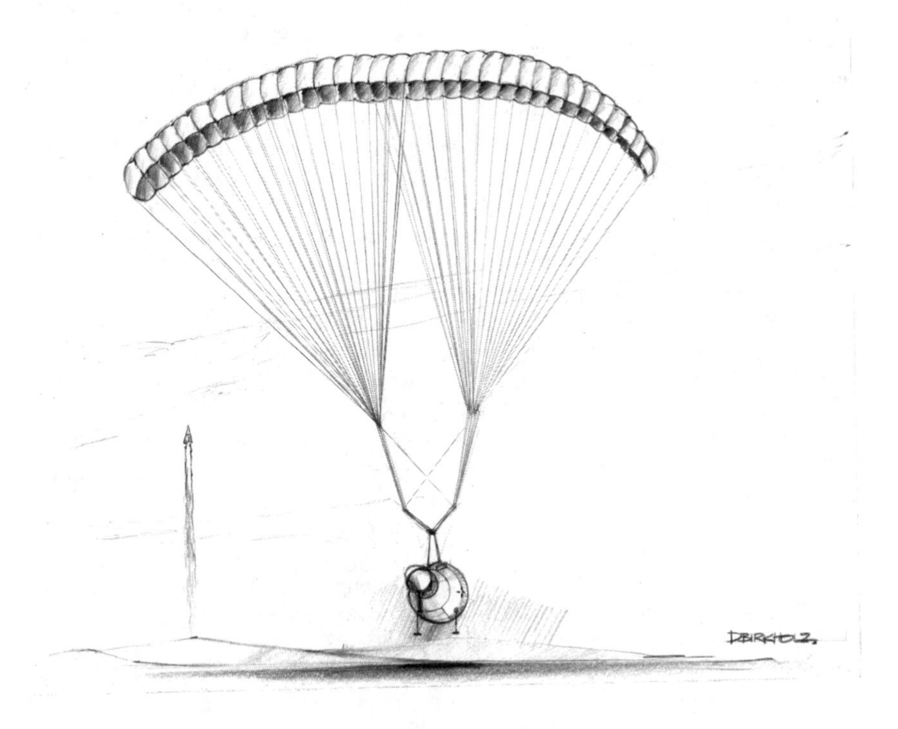

AS part of their systems engineering methodology, the aerospace and defense industries make a big deal of design reviews. These are technical presentations made to the customer—one of the military services or NASA—to apprise the customer of the progress that has been made in a major system development program. Because of the complexity of these programs, and the long periods of time over which they are spread, a program might continue for years without making any substantial progress. The design review was instituted to both spur the contractor on by requiring them to accomplish certain tasks for each of these milestones and to provide a means of ferreting out any serious technical or programmatic problems. Usually at least three reviews are conducted: a preliminary design review, a critical design review, and a final design review. The reviews are not intended to be adversarial; presumably, all involved parties have the same goal in mind: success of the program. In fact, however, they often became adversarial because rival factions within the defense establishment, Congress, or the administration can use the results of a design review to kill a program. They can also turn into forums for one-upmanship, with reviewers more interested in demonstrating to the world their intelligence (and egos) than in understanding where the design actually stands. Likewise, they provide opportunities for organizations other than that of the primary customer to wield influence by asking questions and demanding the addition of features that might have very little to do with meeting the original system requirements.

At AM&M, there was no customer to whom to give a design review. The investors, of course, were closely watching the progress of the vehicle design—and the rate at which their money was being spent. But the investors were not engineering experts, or even knowledgeable enough to question the judgment of Tom and his team, and Forsyth made sure that any questions the investors had were promptly answered. They wandered around the facility from time to time and received regular reports of what was being accomplished, and so they did develop a feel for the people and the organization, in the sense of whether it was functioning well or poorly. At some point, prospective customers for the DH-1 would, like the customers for a commercial transport aircraft, be involved in setting requirements for certain aspects of the design, but they would certainly not be involved in supervising the engineering. No one doubted that Tom was fully in charge of and responsible for the design. He and his top managers were the only people that Forsyth trusted to question any particular design choice or the competency of any particular engineer. Within the company, though, the atmosphere was familial, in the sense that they were all going in the same direction with the same goal in mind: to build a truly low-cost, reusable spacecraft and to make a successful business of selling it.

Of course, as in any family, there were bound to be arguments, and engineers, although not generally adversarial, could be highly opinionated. As chief engineer, Tom either arbitrated—or in some cases encouraged—disputes, as seemed appropriate. Typically, the parties involved would eventually resolve the issue among themselves. But if not, Tom didn't hesitate to play the *pater familias* and make the final decision. Mostly, however, he preferred to gently encourage and inspire the engineers to follow a logical path that was consistent with the overall goals of the project. They had been hired for brains and the ability to get things done, and the level of authority and responsibility with which each was invested was rather sobering, so that they tended naturally to take seriously the judgment and opinions of their peers. It helped that the plant at Cape Canaveral was laid out like a miniature Redstone Arsenal, with test stands, labs, and engineering offices located in a single facility. Problems discovered in test or fabrication could and would be worked on the same day in the labs or by the engineers. Borrowing a page from Hewlett Packard, Tom also instituted the tradition that, whatever your project or task, drawings, sketches, or models should be set out in your office so that others, wandering past by accident or intent, could look at your ideas, play with your models, and provide feedback—both positive and negative.

...whatever your project or task, drawings, sketches, or models should be set out in your office so that others, wandering past by accident or intent, could look at your ideas, play with your models, and provide feedback–both positive and negative.

All of this enabled the individual engineers to be quite familiar with parts of the system other than the ones they worked on directly. Those who had a broader range of interests and capabilities, as long as they were getting their work done, were free to indulge in reviewing, understanding, and even nudging other parts of the project in the right direction. The primary engineering management tools consisted of two computer-based models: one for the prototype and one for the production vehicle. Every attribute of each vehicle was held in a common database, and changes were made by individual engineers according to their responsibilities. The system included live engineers and librarians who reviewed, logged, tracked, and, where appropriate, alerted other engineers when changes were made that affected their systems. Those without authority to make changes in particular areas could nevertheless propose changes, as long as they provided the rationale and were willing to discuss their proposals with the responsible engineers.

Of course, at some point the design had to be frozen, and Tom had decreed that for the prototype that would be at one year after go-ahead. By that time, it was necessary to begin ordering certain types of flight hardware even for the production vehicle—such things as actuators, flight control software, and ground-handling equipment—and to begin preparation of flight-test handbooks that would form the basis for the launch license applications. By the one-year point, a detailed weight budget had been developed for both stages. On the first stage, 70% of vehicle weight had been determined from actual hardware or calculated from build-to CAD models. The remaining 30% was based on engineering estimations, the accuracy of which varied depending on the experience of the engineer and the maturity of the design. The second stage was less well defined, not in terms of the amount of design work that had gone into it, but in terms of the necessary precision that this stage's sensitivity to weight growth demanded. Every engineering group had a weight budget, and Tom was holding to a nominal 4000-pound payload. Nevertheless, he expected to lose most of that as first flight approached. Tom knew that most problems could be solved by adding weight, and if the first flight date began to slip too far he would not hesitate to add whatever weight was necessary to start the flight-test program by the two-and-a-half-year point.

WITH any new space launch vehicle, there was always an inherent credibility gap. All of the computer simulations and all of the large pieces of hardware that could be thumped (but were still sitting on the factory floor) wouldn't convince anybody that the vehicle was going to be for sale any time soon. Forsyth had made it clear that an early and important objective of the program, which might be critical to raising the additional funds they would need, was a successful flight to orbit and recovery of the orbital stage. Certainly private programs were not the only ones that had spent lots of money on launch-vehicle development without ever getting to the flight-test phase; many government programs too had been axed because they never seemed to be getting close to flying anything. At some point, in the absence of successful test flights, neither the capital markets nor Congress show much inclination to provide further funding.

Forsyth's strategy for the program was not unlike the vehicle itself. One thing only was important: low cost—not 20,000 pounds to geotransfer orbit; not 4000 pounds to low polar orbit, where a 50, 40, or even 20% miss on the targeted payload meant missing the market entirely; not automatically deployed payloads, not backward compatibility with existing or already designed payloads; not excellent cross range; not long duration on orbit—none of these things. Just low unit cost and low cost per flight, and demonstrations of those critical attributes were essential for selling the vehicle. The only way to do that was with test flights that provided convincing demonstration of the feasibility of the overall design, and that meant manned flight into orbit, recovery, and repeated reflight of the orbital stage. That was the first major goal.

And if that meant that the first stage had to be expended in the first 5 or 10 flights, then so be it. Tom directed the first-stage engineers to develop a contingency plan that would allow them to increase its propellant load. With its high thrust-to-weight ratio, the first stage could be modified to carry almost another 100,000 pounds of hydrogen peroxide and propane. With more propellant, the DH-1 could be flown along a classic injection trajectory, and the nominal payload of the orbital stage could be more than doubled. This option, though, would probably entail the loss of the first stage because of its substantial downrange velocity and impact into the water. But the prototype first stage had an estimated cost of "only" $10 million, and Tom had three complete stages on order with long-lead items for another three stages—just in case. But they had to get that upper stage to orbit. Even if they had to take this detour, Tom was sure he would be able to get a useful payload out of the first production vehicle and that it would get better from there. Better a detour than a dead end.

Just what would constitute acceptable performance for the first production vehicle was not set in stone. The short-term goals were, for the present, one flight per month with some "useful" payload; Tom was hoping for 5000 pounds, but in truth 2000 pounds or so would be enough. After all, even the space shuttle had had less than half its design payload at first flight. AM&M was going to succeed, in Tom's opinion, but not because they had better technology, or any specific advancements in technology.

In fact, the DH-1 or a vehicle like it could have been built with mid-1960s technology. Some 40 years later, a lot of things were easier, especially the electronics. But, no, it wasn't the technology. The big difference was a change in goal. The goal was not to advance hypersonic airbreathing propulsion, or exotic composite technology, or any technology. No, the goal was to build a reusable vehicle as cheaply, as reliably, and as simply as possible, and to sell that launch vehicle to others who, operating in a competitive environment, would drive the cost of launch down to the point where new markets would come into existence. That would ensure growing demand for the vehicle and its follow-ons. One of John Forsyth's dictums was that the market would initially be driven—like almost all major technical innovations—by faith, hope, and buzz. Long-term success for AM&M, though, would depend on achieving real reusability and low costs.

...the goal was to build a reusable vehicle as cheaply, as reliably, and as simply as possible, and to sell that launch vehicle to others...

THE first step was a successful prototype flight-test program. Tom undertook, before the prototype design freeze was declared, an internal design review. It consisted of an all-day series of lectures and presentations on the various vehicle parts and subsystems so that everyone would be as up to date as possible on the design before they made a major commitment to buying and building hardware. The review also served to force the engineers to go through a checklist of sorts to ensure that they had everything ready that needed to be ready. It also provided them with the opportunity to understand the design solutions and decisions that had been incorporated to make sure that everything was internally consistent and workable.

Of course, some hardware commitments had already been made, at least for the engines, even before the project had an official go-ahead. In addition, engine specifications for the first-stage test vehicle had been frozen at three months, and the basic layout of the pilot's station and a reasonable desktop flight simulator had been settled at six months. And very early on, several hangar queens had been built for both the first and second stages, so that the engineers would have something real to look at as they considered the design details. One of the orbital-stage hangar queens was sufficiently close in weight to the flight prototype to use as a drop-test vehicle with the parawing, which was expected to be available soon. The one-year design freeze was simply one in a series of design decision points.

The review began with the prototype first stage. Unlike the second stage, the tanks were not integrated into the structure. Rather, there was a frame of high-strength 7000-series aluminum onto which the various propellant tanks were mounted. This was done in such a way that, as the tanks stretched elastically under internal pressure, only the mountings had to accommodate the stretch, not the entire airframe. This facilitated the use of separate propellant tanks for the main and landing engines. It also allowed use of relatively simple designs for the engine mounts and the quad-redundant hydraulic actuators.

Prior to launch, the vehicle would be supported by four 5-inch diameter steel posts, topped by gas shock absorbers that were uncharged initially. These posts, fixed to the launch stand, would slide into four "pad supports"—6-inch diameter, 15-foot long vertical tubes, capped at the top and mounted symmetrically inside the first stage around its axial centerline. Upon ignition of the main engines, the vehicle would fly along the four posts for 15 feet, reaching a velocity of about 25 feet per second, and then fly free. It would only take about 1.25 seconds for the vehicle to rise off the posts, but should the engines be shut down within the first half second or so, the gas shock absorbers, which would begin to charge as soon as the weight of the vehicle came off the posts, were there to absorb the vehicle's weight as it settled back down at up to 15 feet per second.

With this system, the large dynamic loads resulting from the sudden release of vehicle hold-downs at full thrust are eliminated. In a typical launch vehicle, sheer plates or explosive bolts hold it down until thrust builds up, with the vehicle pulling on the launch stand. When the rocket is suddenly released, the force of the engines' thrust is no longer resisted by the hold-downs, which results in high transient loads that can become the design case for the vehicle structure. The impact of this phenomenon on the weight of the structure was well documented during the British Blue Streak booster program in the 1960s. The launch stand itself would be of the "milk stool" type, much like the one used for the V-2. This type of stand can be made as a prefabricated weldment and requires only a simple, if beefy, concrete foundation to which it can be bolted. There would also be sound suppression and water deluge systems.

Acoustic loads, which can be quite severe on a large rocket, are generated from the sound from the engines being reflected back onto the vehicle by the pad and surrounding area. They typically reach a maximum just after takeoff, when the vehicle is still close to the ground. Because of the relatively high thrust-to-weight ratio of the first stage, and correspondingly higher acoustic loads, Tom had put a small team to work designing a water-spray sound suppression system that would actually fly alongside the vehicle for a short time.

It was not immediately obvious to me why spraying water around the vehicle would reduce the noise. One of the engineers provided an explanation. Small droplets of water suspended in air change the speed of sound to about 200 feet per second, as opposed to the typical 1000 feet per second in dry air. This change in acoustical impedance has an effect similar to the optical effect caused by a large change of index of refraction, which traps light by causing total internal reflection. Sound waves within the high-density air-water droplet mixture have difficulty leaving and become entrapped. Thus, the sound energy from a launch could be substantially reduced if water sprays could be mounted on the vehicle. Obviously, the weight penalty of carrying water on the first stage for sound suppression would be severe. However, one of Tom's engineers had come up with a scheme to connect a high-pressure hose to the first stage, supported by a steam-driven catapult structure on one side of the vehicle that would ride up along with it for the first 300–500 feet of flight. That would be enough to greatly reduce the sound and the acoustic loads. It was too wild an idea, of course, to try on the first test flights, but Tom had set aside a small amount of money to test its feasibility after the flight-test program—if a functioning first stage were still available. If it worked, over 200 lb of acoustic insulation in the payload bay could be eliminated.

The prototype first stage, 25 feet in diameter and 30 feet high (including the engine nozzles but not the extended landing gear), would be boosted aloft by five pressure-fed propane-hydrogen-peroxide engines, mounted in an X configuration. The center engine would be fixed, but the four outboard engines would be able to gimbal up to 12 degrees in the outboard direction (in engineering parlance, each engine would have a single degree of freedom). Power for engine control would come from quad-redundant hydraulic actuators. The four vernier engines would be mounted on a separate thrust structure positioned just inboard of the vehicle's periphery. These engines were controlled by aircraft-type, quad redundant two-axis electrohydraulic actuators.

Electrical power would be provided by double-redundant lead acid batteries backed up by a small thermal battery. Nominal flight time of the stage was less than 500 seconds, so that with a total power requirement of 20-kW average load and 100-kW peak on a 32-V DC bus with a 400-Hz aircraft-type inverter the total weight of the electrical power and distribution system came to less than 500 pounds and total capacity was less than 3 kW hours.

The engineering staff had suggested that this baseline system could be replaced by a hydrogen-peroxide-powered turbine for a total system weight of approximately 200 pounds, including the peroxide propellant and storage system. However, for the sake of overall system reliability, cost, and complexity, and in view of the relative insensitivity to weight growth of the first stage, especially for the prototype, Tom had decided to stick with the battery system.

The propellant tank system consisted of a single, central propane tank with four hydrogen-peroxide tanks placed symmetrically around it. The propane tank had a large bulb at the top so as to provide more pressure head. This tank would be pressurized by evaporating some of the propane in a heat exchanger on the center engine and then directing it back to the top of the propellant tank through a 4-inch duct. For the peroxide tanks, a separate, regulated high-pressure source of hydrogen peroxide was run through a dedicated catalyst bed to provide oxygen for pressurizing the tanks. Seventy-percent hydrogen peroxide was chosen for this system, in order to keep the temperature of the pressurant gas below 400°F.

The second stage would be mounted onto the first stage by means of four support posts located just around the central engine bay and tied into the main-engine support structure. Stiffness of the mated combination would be enhanced by an array of petal-like devices mounted around the upper rim of the first stage. During first-stage flight, these petals would be closed around the aft end of the orbital stage, imposing an inward compressive load on the sides of hydrogen tank. They would be kept in contact by a series of pneumatic pistons connected to a common, triply redundant gas manifold. The petals were graphite composite panels, approximately 4 feet long and coated with Teflon® on the places where they contacted the upper-stage tank. They were also heated to make sure that no ice or condensed air could freeze the upper stage to the petals. At stage separation, the spring-loaded petals would pop outward and latch into the open position, looking rather like a flower. During first-stage descent, the panels would remain open to increase the aerodynamic drag area by about 30% and also to move the center of pressure well aft (actually upward, in the descent mode) of the center of mass, thus improving stability.

> **The second stage would be mounted onto the first stage by means of four support posts located just around the center engine and tied into the main-engine support structure.**

The vernier engine tanks were sized for 11,000 pounds of hydrogen peroxide for landing plus 1500 pounds for attitude control during the vehicle's climb from main engine cutoff at 100,000 ft to 200,000 ft, where staging would take place. Two independent hydrogen-peroxide attitude control systems were provided for use during the 220-second coast between vernier engine shutdown and reentry into the atmosphere, where aerodynamic forces would again be sufficient to stabilize the vehicle.

Because they were already going to be on the vehicle anyway, Tom had elected to use the four vernier engines for a rocket-powered vertical touchdown of the first stage. Because they are peroxide engines using a catalyst bed for ignition, they could be rapidly and reliably started during the critical final moments of the descent, and they were easily throttled. Drop tests of one of the first-stage hangar queens, with the addition of a small stabilizing drogue chute, had verified small-scale wind-tunnel test results, demonstrating a terminal velocity of about 500 feet per second at an altitude of roughly 5000 feet. This means that when the vehicle reached a 500 feet per second rate of descent, the total aerodynamic drag force equaled the force on the vehicle caused by gravity, that is, its 40,000-pound weight. Thus, the velocity would subsequently remain constant, until the engines were started.

As the first stage was slowed by the upward-thrusting vernier engines, the drag force would fall off as the square of the velocity. Nevertheless, the drag forces would continue to help slow the vehicle down. The four vernier engines would have a total maximum thrust of 60,000 pounds, providing a thrust-to-weight ratio of 1.5 at start of the landing burn and 2.0 at landing when the stage weight would be near 30,000 pounds. Including drag, it would take just over 20 seconds to bring the vehicle to a stop. The maximum weight of propellants required for landing, including enough for 10 seconds of hover before touchdown, could be calculated by taking the maximum thrust of 60,000 pounds divided by the specific impulse—160 seconds—to get 375 pounds per second for 30 seconds. That comes to about 11,000 pounds of propellant. It would come to rest on four deployable, shock-absorbing landing gear mounted outboard around the base of the stage.

The flight-test plan involved flying the first stage initially with the vernier engines alone, the main engines being replaced by equivalent-weight dummies. This would allow the landing systems to be thoroughly tested and provide an opportunity for the pilots to practice the terminal landing phase. Following the manned test flights with the verniers alone and captive-test firings with the main engines, the first stage would be flown on a typical flight profile, carrying only a dummy stage in place of the second stage. The main engines would be throttled back to compensate for the lighter gross weight of the vehicle. The total delta-V that could be achieved with the first stage had to be kept to no more than 3000 feet per second, otherwise, the reentry heating design levels would be exceeded.

The prototype orbital stage, being more weight sensitive, would have to be somewhat closer to the production vehicle configuration. Three flight-weight structures were being constructed, although the third one would be held back and finished only if one of the other two prototype vehicles were lost in test. The hydrogen and oxygen tanks were assembled from individual segments that were rolled, stir-welded together, and then integrated with the interstage structure. The engines were more or less off the shelf, although the baseline of six RL10 engines had been changed to two slightly modified RL60 engines. The by now flight-proven RL60 engines brought the weight down and increased performance. Pratt & Whitney had been able to squeeze six of them out of their current production run, as a result of delays in other launch-vehicle programs caused by a program cancellation and by one launch accident that led to the inevitable stretch-out of two other flights.

FOR the reentry thermal protection system, Tom took a very conservative approach. The first vehicle would have a 1500-pound ceramic blanket heat shield made with specially designed fibers and filled with a proprietary compound containing 1.5 pounds of water of hydration per square foot. The second prototype would use a more advanced blanket material with only a half a pound of water per square foot. This configuration was expected to keep the surface temperature below 1500°F, but would be capable of withstanding 2700°F with the ceramic alone. To protect the engine bay, the first test vehicle would deploy an umbrella-like heat-shield blanket to cover the engine opening. The second was to use a cover deployed by a coiled metal ring, which, after deployment, would be drawn tightly against the vehicle base by four ceramic ropes. To prevent damage caused by any hot gas penetrating the engine bay, both vehicles would have a water spray system that would keep the bay continuously filled with a mist during peak reentry heating. The system consisted of a lightweight, composite blowdown tank containing 75 pounds of water that would be delivered through an array of spray nozzles.

The exterior of the vehicle forward of the base heat shield was covered with a thin nomex blanket, underlain where necessary by a layer of lightweight quartz felt filled with the hydrated, or water-bearing, compound. The amount of hydrate was tailored to the location on the vehicle, according to predicted heating loads. This system yielded an average weight of 0.25 pounds per square foot for the blanket, felt, and hydrate for a total of between 400 and 500 pounds for the heat shield on the conical sides of the vehicle. A ring of solar cells around the payload bay, which constituted the on-orbit power system, was backed by a phase-change heat sink consisting of about 0.5 pounds per square foot of lithium hydride, which could absorb 550 Btu as it melted. This was separated from the payload bay wall by a thin ceramic blanket.

The landing system comprised two redundant parawings, mounted on either side of the vehicle centerline in the interstage. Each parawing was stowed in a separate compartment with externally opening, electrically operated doors. The doors could also be opened by means of a ripcord mechanism manually activated by the pilot. As additional backup, a mortar-deployed round parachute was contained in the nose of the vehicle. Each parawing had a drogue chute, one of which would be deployed to stabilize the vehicle in a base-down attitude when the attitude control jets became ineffective at about 100,000 feet. At about 30,000 feet, additional cable would be released, bringing the vehicle to a horizontal attitude for deployment of the parawing. In the event that the first parawing failed, it could be cut away, and the second could be deployed. If there were problems with the second parawing, it too could be cut away by explosive actuators, and the pilot could deploy the round chute, which was possible down to an altitude of 200 feet, or bail out using the ejection seat.

The landing system comprised two redundant parawings, mounted on either side of the vehicle centerline in the interstage.

The main landing gear would consist of two dual-wheeled units that retracted into a pair of cylindrical bays tied into the hydrogen tank structure where the sidewalls were joined to the base. These bays were located about 1 foot up from what would be the bottom of the vehicle in its landing attitude and 14 feet apart. During reentry, they would be protected by beryllium heat-sink doors that were mounted to the gear. The front gear would be stowed under the epoxy-graphite nose cone at the top of the vehicle, which would be extended forward by a pneumatic strut to allow the gear to drop down into position for landing.

Using a mock-up of the orbital stage, drop-and-deploy tests of the landing systems were scheduled for the next 6 months, leading to final design decisions at 18 months into the prototype program. This was the only series of flight tests for the upper stage, short of flight into orbit, that was deemed sufficiently low risk. Even though Tom was a great believer in gradually expanding the envelope of a new flying vehicle with incremental flight testing, there were no reasonable options between the drop tests and full-up flight to orbit. Recovering the second stage from a partially fueled, suborbital flight could actually present more severe reentry conditions than those from orbit, and although the vehicle was designed to survive them in the event of an in-flight emergency, Tom would rely on simulations and the orbital test flights to verify the suborbital abort modes.

AM&M
AM&M

CHAPTER 15
Guidance, Navigation, and Control

AFTER hearing Tom describe the flight-test program, I was all fired up to take a hop myself. A detailed mock-up of the first-stage pilot station, including a flight simulator, had been built in the back of one of the engineering test labs, and so I went over to see if I could get a "ride." Len Donovan, the test engineer in charge, led me over to the mock-up. I immediately noticed what looked like a jet fighter canopy—something I didn't expect to see. I asked Len about it.

"Good call," he said. "That's actually a canopy from a Russian MiG-29S. For the prototype, it's an off-the-shelf solution that will work just fine. It's got a built-in jettison capability, and it's easy to integrate with the pilot's ejection seat, which also comes from the same airplane—with some modifications. Both are relatively simple, very reliable, and the price was right." He slid the canopy back, and I peered inside. Instead of a position that allowed the pilot to look straight out away from the vehicle and down to where the ground would be, the seat orientation put the pilot flat on his back.

"Won't that make it harder for the pilot to land it?" I asked. "As I recall, the Rotary Roton ATV had the pilot sitting upright."

Len replied, "Our first idea was to put the pilot on his stomach, which would be a more natural position for a vehicle that lands vertically. But that would have required development of a completely new ejection seat, so we dropped that idea early on. I think you'll see, though, that we came up with a nifty way to make it work. Anyway, rocket pilots should be on their backs at lift-off—and remember, this baby has a pretty high thrust-to-weight." With that, Len motioned me inside.

Climbing into the cockpit, I saw three liquid crystal displays in front of me and something that looked like an articulated telescope to the right. There was a panel of switches and a joystick on the right side, as well. Below the liquid crystal display (LCD) screens was a specialized keyboard. There was another joystick on the left-hand side.

I settled into the seat as Len said, "Since the trajectory of the first stage is simplicity itself—straight up, straight down—the pilot really doesn't need to see anything other than the landing pad below him. No maneuvering, no worries about other traffic, at least at the Cape here. The central screen shows what's directly underneath by means of a triply redundant video camera system. Pointing to the telescope-like device, he said, "This allows you to look down below through a purely optical system. But the best way to see how everything works is to fly it! Here, fasten your harness." So with the canopy open, Len reached in and punched some keys to reset the system, and the screen began to give a 30-second countdown. Len said, "To start out, just leave the stick in the middle—don't touch it. After you gain some altitude, you can see if you're drifting away from directly above the landing pad, and then use the stick to make corrections with the thrust vector control system."

When the count reached zero, I was a little startled to feel the seat give me a jolt. Len explained that a small amount of motion simulation had been added to demarcate the various phases of the flight. As I watched the screen, the "ground" zoomed away from beneath the vehicle; it seemed a lot faster than I had seen it on the webcams mounted on other, conventional launch vehicles. It even seemed faster than the shuttle. Len pointed to the telescope-like apparatus and said, "Take a look through the scope." I pulled it to my eye and saw a similar image of the ground receding into the distance. I could now see that the landing site was marked like a giant bull's eye, each ring a different color, with large numerals prominent on each ring. Very soon the entire bull's eye was visible. Len told me it was only half a mile in diameter. Then I noticed, radiating outward from the cardinal points around the bull's eye, rows of bright lights. I asked about the lights, and Len said, "Those are collimated lasers, targeting the vehicle from one mile, two miles, and three miles out. Check your position now." On the screen in front of me a crosshair had started to drift off center. I reached for the joystick and with a few small movements easily brought it back to center with only a little overshoot. Looking through the eyepiece again, I noticed that the bull's eye was not staying centered in the crosshairs, but rather was moving along what appeared to be an illuminated ellipse that overlay the crosshairs.

"Why is the bull's eye moving?" I asked.

"Don't forget the Earth is rotating under you," Len replied. "Even though the trajectory is basically straight up and down, you still have to fly back a bit to the landing site. The camera and screen system compensates for that little orbit, so that you can steer by keeping the crosshairs on the bull's eye. The optical system doesn't."

I watched the central screen for the rest of the flight, making minor adjustments with the stick from time to time. The seat gave a slight rumble as I passed through maximum aerodynamic pressure, or "max Q," and the sound level (which Len had informed me had been turned way down for demonstration runs) began to drop off noticeably. The simulator varied the noise as the atmosphere would thin out during the ascent, including adjusting the pitch to account for the changes in the sound of the engines as transmitted through the vehicle structure. When the altimeter reached the 100,000-foot mark, the seat gave a jerk corresponding to main engine cutoff. Because I was flying a "production vehicle trajectory," that meant a thrust level of some 70,000 pounds. The readouts indicated continued upward flight, but with the velocity gradually falling off, as a result of the vernier thrust to weight of less than one. At 200,000 feet, it was time for staging. Again, I felt it through the seat and watched on the left-hand screen as my simulated orbital stage separated and pulled away. Then the seat gave another jerk denoting vernier engine shutdown, and the sound died away.

The first stage continued its simulated climb, coasting up to around 350,000 feet. Above staging altitude, the vehicle was now responsive to the reaction control system, which was controlled by the same joystick on my right. Without significant aerodynamic forces acting on the vehicle, control was even easier, and it took very little motion of the stick to keep the crosshairs on the bull's eye. Soon I was descending, and I looked at the right-hand screen, which showed a view over my shoulder of the Earth's limb. It looked pretty nice, even if it was only a simulation. At about 140,000 feet, the thin whistling of air over the vehicle became audible, and the reaction control system became sluggish. I looked at Len with a what-do-I-do-now look, and he pointed to a switch on the main panel. "Flip the switch marked 'DROGUE.' " I did so and heard a loud snap. The left-hand screen now showed a drogue chute deployed above the stage, which stabilized my descent.

In the center screen, the three-mile laser lights hit the edge of the sighting circle, indicating 50,000 feet altitude. I glanced at the altimeter and saw that it read 50,000 also. Len pointed to the lower right corner of the screen, which displayed a rapidly increasing number. "That's your predicted miss distance from the center of the bull's eye," he said, "Grab the left stick now. You need to fly the drogue." On the screen, a line appeared between the center of the display screen and the bull's eye on the ground, providing a graphic indicator of how far off I was. With the left-hand stick, I was able to control the shrouds to the drogue and maneuver enough to bring my drift to a stop and come—slowly—back, so that I was again centered over the bull's eye. At this point an alarm sounded, indicating that the stage was below 12,000 feet, and the center screen began a countdown to landing engine ignition at 7500 feet. Len said, "Get your right hand on the throttle. Once the landing engines ignite, you have a little more than 30 seconds of fuel—but 20 seconds should be all you need for a perfect touchdown." Just as the one-mile lasers indicated that I was at 7500 feet, the seat gave another jerk, and I heard the rumble of the landing engines. My rate of descent began to slow immediately. I had played "Lunar Lander" as a boy, and I knew that the trick was to not lose

I had played "Lunar Lander" as a boy, and I knew that the trick was to not lose too much velocity too fast, or you run out of fuel while you're still way off the ground.

too much velocity too fast, or you run out of fuel while you're still way off the ground. But it had been a while. I concluded my flight with a "controlled descent into terrain" at 100 feet per second, and was rewarded with an ejection seat deployment warning just before impact.

After the screens cleared, I asked Len about the telescope. "I didn't really need it. Is it a backup?"

He grunted and said, "Yeah, it's a backup. To tell you the truth, at first I thought it was just a silly anachronism that Tom had insisted on. But later, I found out that the pilots really like it. Whether it's because it offers a little more challenging way of flying the stage, or because they like the idea of having a purely optical-mechanical backup system for control, I'm not sure."

"How exactly does it work?" I asked.

"Well, it is basically a telescope," Len began, "and it gives you a view directly beneath the vehicle. The 'squashed ellipse' nominal path that the vehicle should take is illuminated within the eyepiece. The tick marks you saw indicate where the vehicle should be at various altitudes throughout the flight. Altitude is mechanically calculated by bringing two mechanical indicators or pointers to the sides of one of the landmarks on the bull's eye. There are six adjustable ranges—one for each ring of the bull's eyes between the inner bull's eye and the outermost lights, and by selecting a range and twiddling that knob on the left there, you get a mechanical/digital instrument read-out of your altitude in feet. That can then be checked with the nearest tick mark, which should be over the center of the bull's eye. If it's not, you just use the stick to bring the vehicle back to where it ought to be. It works the same on the way down."

Len went on to explain how, for all-weather operations, an infrared converter could be added to the optical train, so that infrared lasers could be used through heavy haze or clouds. As another option, the screen-based system could use infrared cameras for imaging with a radar system to provide ranging by means of corner reflectors positioned along radials outward from the bull's eye.

I asked Len, "But why include a purely mechanical system? Don't you need a pretty sophisticated computer just to control the engines anyway? Do you gain that much more reliability?"

Len started to explain, thought better of it, and said, "You should go talk to the GNC people and get a complete rundown on the philosophy behind the vehicle control system, for both stages."

I thanked Len for my "flight" and headed out to see the guidance, navigation, and control (GNC) group up on the fourth floor of the engineering building. As I walked up the stairs—about the only exercise I get these days—it occurred to me that it was appropriate that they were on the top floor. However much their tendency to blow up made rocket engines seem to be at the heart of rocket science, those awe-inspiring engines still had some connections with everyday devices like furnaces, pumps, and automobile engines—all the way up to jet engines, which, even if you don't fully grasp the engineering that goes into them, at least seem to be accessible to understanding. Not so with guidance, navigation, and control.

Media commentators are always describing spaceflight guidance and navigation problems as akin to hitting a dime with a rifle from so many miles away while both you and the dime are moving, but that doesn't leave you with a very firm feeling for how difficult the problem really is, or even whether it can be understood and solved by mere mortals. I got to the fourth floor only a little out of breath, and walked down the hall to the door labeled Guidance, Navigation, and Control.

There were only four people in the GNC group: chief Tim Hansman, Phil Kiffler, John Kanellos, and Paul Reston. In keeping with the general scale of things at AM&M, all of the teams were small, worked long hours, and let the contractors build the hardware and write the software to their specifications. Actually, of these four, only three were card-carrying GNC designers. Paul, trim and fit but obviously the oldest of the group, was introduced by Tim as a performance man. When I asked him what that meant, Paul laughed and said, "I tell these guys when they screw up."

Tim said, "That's the way he likes to put it, but yes, Paul keeps us on the straight and narrow. It's his job to provide a good, intuitive understanding of the whole problem and come up with quick, first-cut answers so we or the contractors don't go off in a wrong direction—pun intended."

As we talked, I could sense that Paul was an independent spirit who, as he told me, loved to have the first crack at all the problems single handedly, using computer programs he had written himself, dealing with all aspects of trajectory design, flight control, and vehicle performance analysis. In any technical field, there is always the danger that the experts will come up with the right answer to the wrong problem, which they had set for themselves but was totally divorced from the problem they actually needed to solve. Paul knew what the right problems were, and roughly what the answers should look like. By reason of Paul's all-encompassing insight and intuitive grasp of what GNC is all about, I found that he was the best one to explain to me what exactly Tim and his team were doing and how the DH-1 was going to be flown. I knew I would never be able to wade through the mathematics that the other three were perfectly happy to wallow in all day long: Hamiltonian dynamics, integrating around holes and poles on the complex number plane, and worse. I asked Paul to give me a layman's overview of GNC.

In any technical field, there is always the danger that the experts will come up with the right answer to the wrong problem…

"Well," he began, "what it boils down to is knowing where you are and where you want to go—that's navigation; figuring out how to get from where you are to where you want to be—that's guidance; and then being able to steer your vehicle to take you there—that's control."

"Sounds simple," I said, "but I know there's a lot more to it. How do you do those things in a rocket?"

Paul started with navigation: you have to know where you are before you can successfully get to anywhere else. Typically, in a rocket or missile, you use a so-called inertial reference platform to tell you where you are. In the early days, this consisted of a platform suspended within three concentric gimbals, so that the platform could rotate about the three Cartesian coordinate axes. Motion about each axis was effected by a servomotor so that the platform was maintained in the same orientation relative to an inertial reference frame at all times. (An inertial reference frame is a coordinate system that is fixed relative to the distant stars.) Mounted on the platform were three accelerometers—each basically a weight on a spring, with damping to take out vibrations—which continuously measured the acceleration in all three coordinate directions. These accelerations can be integrated once to give the velocity in each direction, and the velocities can be integrated to give the distance traveled in each direction in a given time interval. Thus, the output of the three accelerometers can be used to tell you where you are relative to your starting point, as well as how fast you are going.

There is something else you need to know besides where you are, and that is where you are pointing. This is determined using rate sensors. The simplest form uses a spinning gyroscope or rate gyro, which, because of its angular momentum, will precess, that is, its axis will rotate about its original axis, when a torque caused by the angular velocity of the vehicle as it turns is exerted perpendicular to that axis. The motion of the gyroscope's axis causes it to pull against a spring such that the deflection is directly proportional to the vehicle's rate of turning. Three such rate sensors will measure the rates at which the vehicle is turning about each axis. The output of the rate gyros is used to keep the inertial reference platform fixed in orientation. Integrating the output of the rate gyros will then give you the orientation of the vehicle in the inertial reference frame, again relative to its initial orientation. For the first generations of ballistic missiles, the accuracy with which they could deliver their warheads was dependent on just how well you could build the highly complex, electromechanical inertial platforms, and it was definitely top secret stuff. Nowadays, however, they have been superceded by so-called "strapdown" platforms, in which all of the sensing devices—accelerometers and rate sensors—are solid state components, and microprocessors are used to untangle the accelerations about each axis, so that a stabilized platform is no longer required.

Along with elimination of the need for the high-precision mechanical guidance platforms came dramatic increases in accuracy and reductions in weight. Paul commented that although Tom liked to maintain that the technology necessary for building the DH-1 had been available since the mid-1960s, inertial navigation systems had gotten a lot lighter, a lot cheaper, and a lot more accurate. That's the good news. The bad news is that you still have the problem that the system tells you—with all that improved accuracy—the accelerations and motions of the part of the vehicle to which it is mounted.

All rockets and missiles flex and vibrate to some extent as they fly, sometimes with a frequency as low as 4, or even 2 Hz (or cycles per second). This means that the part of the vehicle where the navigation package is located will experience accelerations and motions different from what the vehicle as a whole is experiencing. One solution is to try to find a part of the vehicle that, because of the structural arrangement, tends to stay motionless with respect to the center of gravity of the vehicle while the other parts bend and flex. However, it is difficult to accurately predict where such a place might be—if one exists at all—because the mass properties of the vehicle are constantly changing as a large part of that mass is expelled through the engines, and the aerodynamic lift and drag loads are constantly changing as velocity increases and the atmosphere thins out.

This is particularly vexing for complicated vehicles such as the Titan 34B, which has two rigid solid rocket boosters strapped onto an assembly of flexible liquid-propellant tanks, or the space shuttle, with four major structural elements pinned together, each with its own bending modes and frequencies. A more practical solution is to place several inertial platforms at various points on the structure, and, through a combination of mathematical modeling and flight data—assuming you survive the first flight, determine the average motion of the vehicle from the inputs of all the platforms. "Anyway," Paul said, "we're starting to get into the problems of control, which I will explain in a minute."

To enhance the accuracy achievable with the earlier inertial navigation systems, space vehicles or missile reentry vehicles used a high-tech form of celestial navigation by taking star sightings, or stellar updates, once the vehicle reached outer space where the stars are always visible. These provided a highly accurate means of updating the attitude, or orientation, of the vehicle as determined by the inertial system up to that point. However, with the level of accuracy that modern inertial platforms can provide, a reasonable job of getting into the correct orbit can be done solely with inertial navigation. Besides, even though it wasn't designed for the speeds and altitudes achieved by launch vehicles, the global positioning system (GPS) can be easily adapted to provide very accurate navigational updates for precise insertion into low Earth orbit.

Once you know where you are, you need to know where your destination is. For flight into orbit, the endpoint is an altitude and a velocity that will allow you to continue around the Earth in a stable orbit. Of course, the problem gets a little more complicated if you want to meet up with a space station, a refueling depot, or another spacecraft. Even more interesting problems arise if you want to arrive over a particular landing site at a particular time. So you also need to predict where you are going to be at various times in the future. Figuring out where you or your target will be in space—as a function of time—requires the application of astrodynamics. Some 300 years ago, Kepler and Newton invented the science of astrodynamics by developing the equations for predicting a body's future position in space given its present position, velocity, and the "orbital elements," which describe its orbit around a sun or planet. However, for real-life problems—where you must take into account more than one source of gravitational attraction, and the fact that the planet about which you are orbiting is not uniform in density or spherical in shape, and the effects of nongravitational forces such as atmospheric drag (even at orbital altitudes), the solar wind, magnetic fields, and light pressure—knowing where your target will be days or even hours in the future is complicated.

The solution is to start with a Keplerian orbit and then constantly add small perturbations, or adjustments, for the other factors. Modern computer power makes it possible to take all of these into account, and reliable orbit predictions can be made hours, days, or even weeks into the future. However, if sunspot activity is high and you or your target is in low Earth orbit, the rise and fall of the atmosphere as a result of solar-wind effects imposes some limitations on orbital prediction accuracy. For flight to another planet, moon, or asteroid, some of these factors become insignificant, but the effect of other bodies becomes correspondingly more important.

I stopped Paul at that point and said, "Certainly the DH-1 isn't going to be required to navigate to the moon, asteroids, or planets." I had heard some of the engineers mention such a thing before, but I never took it seriously.

But Paul replied, "There is a very real likelihood that the vehicle—with some modifications—will be used in such applications. Wasn't it Heinlein who said that once you're in orbit, you're halfway to anywhere?"

"How so?" I asked. Getting to orbit was one thing, but interplanetary travel with the DH-1 seemed like a stretch.

"Because with an additional velocity increment, or delta-V, of less than 25,000 feet per second—which your vehicle has to be capable of to get to orbit in the first place—you can basically go anywhere in the solar system, using on-orbit refueling."

Paul explained that the basic inertial systems and navigational algorithms that are needed for operations in low Earth orbit can be readily adapted for use throughout the entire solar system, simply by including the orbital elements of everything else we know about. And after all, to do really accurate work in orbit, you need to take into account the positions and gravitational effects of the sun, the moon, Venus, Mars, and Jupiter. So if you're keeping track of all of that anyway, keeping track of the rest of the planets, their moons, and the major asteroids really isn't such a big deal. Paul told me that Tim's team was working with a small software company to develop more efficient algorithms for interplanetary navigation, in a program that was implemented on 12 specially configured parallel processors, under the direction of a single Intel processor running the latest version of Microsoft Windows.

Paul continued, "Next comes guidance, or figuring out the best way to get from where you are to where you want to be. And that boils down to designing a trajectory, which can be a path from the ground to orbit, or a path from one Earth orbit to another, or from Earth's orbit around the sun to the orbit of another planet." Paul told how, in the early days of orbital flight, it took a large computer and many, many hours to calculate a single trajectory. No way you could do that onboard. So they precalculated a trajectory and loaded it into the rocket's guidance computer; the vehicle control system then steered along that path, until it reached the endpoint, wherever that was.

There were two problems with this approach. One, if because of wind gusts, or thrust anomalies, or some other unpredicted event you got off track, you might be off so far that you couldn't get back to the correct path and you would never end up at your destination. Second, it was not the most efficient way because, once you got off of the desired path, the best way to get from the new path to the endpoint would be to calculate a new trajectory, something that couldn't be done onboard the old rockets. "Fortunately," as Paul explained, "modern computers—based on processors like the custom chips developed for the interplanetary navigation problem—make it possible to continuously calculate in real time the best path between where you are now and where you want to be. Thus, the guidance computer in the DH-1 would be constantly adjusting or optimizing the trajectory as it flew."

"Now," Paul's voice rose in crescendo, "comes the fun part! Navigation and guidance—that's just mathematics. Now you gotta figure out how to steer that sucker! That's control, and that's a job for real engineers." He told me that virtually all missiles and launch vehicles are inherently unstable, aerodynamically speaking, Burt Rutan's SpaceShipOne being the exception that proved the rule. This means that if you don't constantly fly the vehicle by adjusting the direction of the thrust vector of the engines the vehicle will tip over, and aerodynamic forces will rip it apart.

"Like an airplane that requires that the pilot always keep his hand on the stick," I suggested.

"A better analogy for a rocket," said Paul, "is trying to balance a broom on the end of your finger. With the difference, though, that for a rocket what you actually need to do is keep the vehicle at zero angle of attack—pointing its nose straight into the relative wind—in order for it to remain stable and under control, as opposed to merely keeping it vertical. In a vacuum, of course as long as the thrust vector is pointing through the center of mass, you won't get any moments or rotation of the vehicle. For flight in the atmosphere, though, the analogy of a broom on the tip of your finger is pretty good."

"A better analogy for a rocket," said Paul, "is trying to balance a broom on the end of your finger."

Paul elaborated, "If you hold the end of your finger still, the broom will fall over immediately. But by constantly maneuvering your finger, it becomes possible to keep the broom more or less vertical. Only with rockets, the problem is complicated by the fact that the rocket is constantly flexing. It's as if the broom has a long, really thin handle, and someone nearby is playing rock music with the bass turned up so high that the broom vibrates like a guitar string. Now you have to move your finger, not according to where the broom appears to be at any given moment, but where it will be on average as it vibrates. To make it even more interesting, the time constant of the disturbances—which dictates how much time you have to move your finger to keep the broom from falling, or gimbal the rocket's engine to keep the vehicle from tumbling out of control—are on the order of one-tenth to two-tenths of a second. Actually, it's worse than that—you have to be sure not to jerk your finger around in time with the music, or you will cause the existing vibrations to increase in severity until the broom finally snaps in half." Finally, to make the analogy even more accurate Paul told me to imagine that, instead of having it on the bottom of the broom, my finger is actually connected to a number of springs that press on the bottom of the broom.

At this point, I stopped Paul. "Whoa," I said. "I feel like I do after hearing about shooting a dime umpteen miles away while I'm rotating and the dime is rotating. Can you just tell me what this means for a rocket control system?"

He paused, then went on. "Ok. The basic task of the control system is to take an error signal, derived by comparing where the vehicle is and where it is going with where it is supposed to be and supposed to be going, amplify that signal, and use it to drive the actuators which steer the vehicle, usually by gimbaling the engines to move the thrust vector. In order to avoid exciting the bending frequencies of the vehicle, the control system looks at the output of the guidance and navigation platforms in the frequency domain, and filters out or ignores those frequencies at which it is known the vehicle vibrates. If the vehicle has only a few low-frequency vibration modes that are well characterized, you'll still have sufficient control bandwidth after filtering to control the vehicle. If the frequencies are too low or too many, then you have to use the control system to actively damp out the vibrations. Of course, you also have to take into account the play and flexibility in the actuators and the engine mounts, as well as the mass properties of the engine itself.

"To properly design actuators with the required speed, responsiveness, and stiffness, you need high-fidelity mass properties, aerodynamic, and structural models of the entire vehicle, along with the control laws. Fortunately, automatic control of complex systems is a highly developed discipline, with applications in a lot of industries. There are many excellent, off-the-shelf computer programs that we can choose from to use in designing and building the control system for the DH-1." Paul settled back in his chair.

"In light of all that," I said to Paul, "how could you possibly fly the vehicle to orbit without electrical power? You need some powerful computers going all the time, it seems." I had heard about this capability during the design review, but I hadn't gotten to hear the details.

"Well," Paul said, "If we go back to the basic problem of control, remember that Goddard's rockets, and even the V-2, used simple gyroscopes to control the vehicle flight path. The movement of a gyroscope in response to the motion of a rocket can be directly amplified and used for control, if the rocket is stiff enough and the trajectory is simple enough. Of course, the simplicity of its guidance system was the reason the V-2 was so notoriously inaccurate. But, for our vehicle, the pop-up trajectory is as simple as they come—the first stage flies nearly vertically all the way up and all the way down, and the second stage flies nearly horizontally all the way to orbit. If you can overcome the basic aerodynamic instability of the vehicle, getting the upper stage into orbit can be as simple as blasting the first stage straight up until it runs out of fuel, and then flying the orbital stage along a great circle route parallel to the Earth's surface. A little bit of experience in the flight simulator, a couple of stellar updates once you're out of the atmosphere, and, voila, you're in orbit.

"However, the showstopper is that no human operator can stay ahead of an instability with a time constant of a tenth of a second. And there isn't any way around the instability like the one Sikorsky discovered for the helicopter. Have you ever heard that story? Back in the late 1930s or early 1940s, the government built a small-scale model helicopter and discovered that it was unstable, and so concluded that helicopters were impossible. Igor Sikorsky, however, built a full-scale model, and discovered that while, yes, it was unstable, the time constant was several seconds, which gave a skilled operator enough time to control the thing. The lesson there is that if you want to build something new, better build it full scale so you don't spend all your time solving problems that don't apply to what you really want to build. That's why our prototype is basically the same mass, thrust, and size as the production vehicle.

"Again, though," Paul continued, "there's just no way a human operator can control any reasonably designed launch vehicle. But there is a nonelectrical way of doing it, by using fluidic controls. Fluidics was developed during the 1960s, and at one time seemed like it might beat out computer controllers for industrial applications. It didn't, but you still find 'em in gas flow applications. Fluidics is based on the discovery that you can use small-scale fluid flow to perform switching and amplification functions, much like a transistor. It's even possible to do a limited amount of filtering on gain amplifiers. You can also use gas flow for rate sensors and accelerometers. Building a fluidic circuit is a lot like laying out an integrated circuit in silicon. Individual layers of stainless steel are etched to form the different components and the passageways between them. Then they are stacked up and brazed together, forming a microfluiddynamic device.

"For a launch vehicle, you can build one to perform the functions of a guidance and navigation platform, and use it with pneumatic actuators to drive the hydraulic actuators for engine gimbaling. So that's what we did. The first stage has a three-way redundant strapdown inertial platform and three redundant guidance computers, which used a distributed Ethernet-based network running on both first and second stages of the vehicle and the ground equipment. The three redundant guidance computers drive three redundant sets of actuators. But a fourth set of actuators is controlled by a fluidic platform, and in the event you lose electrical power completely, say, after a lightning strike, you can maintain control of the vehicle.

"Likewise, the orbital stage has three inertial platforms and one fluidic platform. Actually, for the DH-1, using a fluidic guidance platform was rather easy, given the relatively squat shape of the vehicle. That and the pressurization of the propellant tanks, and especially the payload bay, all combine to make it pretty stiff, so a fluidic platform doesn't need much filtering. The most troublesome, most flexible part of any launch vehicle structure—and ours was no different—is the interstage section. To make ours stiffer, John here came up with the idea for the interstage petals. You know, they are locked into position until staging, but when the hydrogen tank is loaded and pressurized, it grows a half an inch in diameter, which even more tightly compresses the tank wall against the petals. Pretty neat idea, but it led to more than a little debate between the structures guys and John and Tim about how stiff it had to be versus how stiff it could be."

After Paul finished, I was feeling a bit dazed, and still unsure about all this GNC stuff, but at least I had a feeling that the AM&M team knew what they were doing. I thanked Paul and Tim for their time, and headed down to try my hand at "flying" the orbital stage.

Phil Cohen was in charge of the orbital stage simulator. The pilot stations of the two stages were very similar, and so it didn't take him long to get me settled in, oriented, and on my way to orbit. Of course, this flight was a little longer than the first, and I just sat there watching the screens for the first 140 seconds of first-stage flight. After separation, Phil walked me through a stellar update with a reference star and the horizon, and without any further difficulty I flew the vehicle into the proper orbit. Phil pointed out that, although I had controlled the vehicle attitude by optical means, the onboard computer had used the attitude information from the optical sight to actually steer the vehicle. If all computing power were lost, it would be necessary to continuously adjust the vehicle attitude relative to the horizon and the reference star. Phil demonstrated how an image of a particular star could be displayed in the eyepiece, with tick marks to guide the pilot to a rough orbit. He also showed me one of those new handheld digital assistants with GPS and interactive voice capabilities. This one had been programmed to independently determine vehicle position and then read out instructions for vehicle pointing. It was fascinating, but I asked Phil why, if a $1000 over-the-counter PDA could be programmed to provide instructions for steering the vehicle into orbit, wouldn't you just use an onboard computer? Why add the purely manual backup system?

Phil explained by referring to the design philosophy that Tom had instilled in the entire engineering team. By making the vehicle at its most basic level extremely

> By making the vehicle at its most basic level extremely simple, you make it easy to incorporate any newly developed, add-on technology such as that embodied in the PDA.

simple, you make it easy to incorporate any newly developed, add-on technology such as that embodied in the PDA. Both stages incorporated a control architecture that included a simple manual capability for control of vehicle attitude. Optomechanical systems provided control capability sufficient for ensuring vehicle safety over the entire trajectory. If the manual backup systems weren't there, you were not too likely to stick $1000 handheld computer into a quarter-billion-dollar space vehicle, even if you could plug it right in.

On the other hand, if you have a robust backup system, you can feel secure using such a simple device to greatly improve the accuracy of the manual system. Phil told me that with the manual system it was also possible to orient the vehicle in space using the attitude control system, and even to restart the engines for the deorbit burn, with a little help from the handheld or radio contact with the ground or precalculated tables in the flight manual.

"Can I go ahead with deorbit and landing?" I asked. "I've been anxious to fly the parawing."

"Sorry, not today," Phil replied. "We just got some new data from the parawing flight test team, and we're still working some bugs out of the landing portion of the simulator. Come back next week and we should be up and running."

I told Phil I would be back and thanked him for the ride. It had been a busy day, but I had had a lot of fun and learned a lot. I was starting to think that maybe even I could become a DH-1 pilot. Time to find out what was going on in the crew systems group.

CHAPTER 16
Webb Suit, Hard Suit, Space Suit

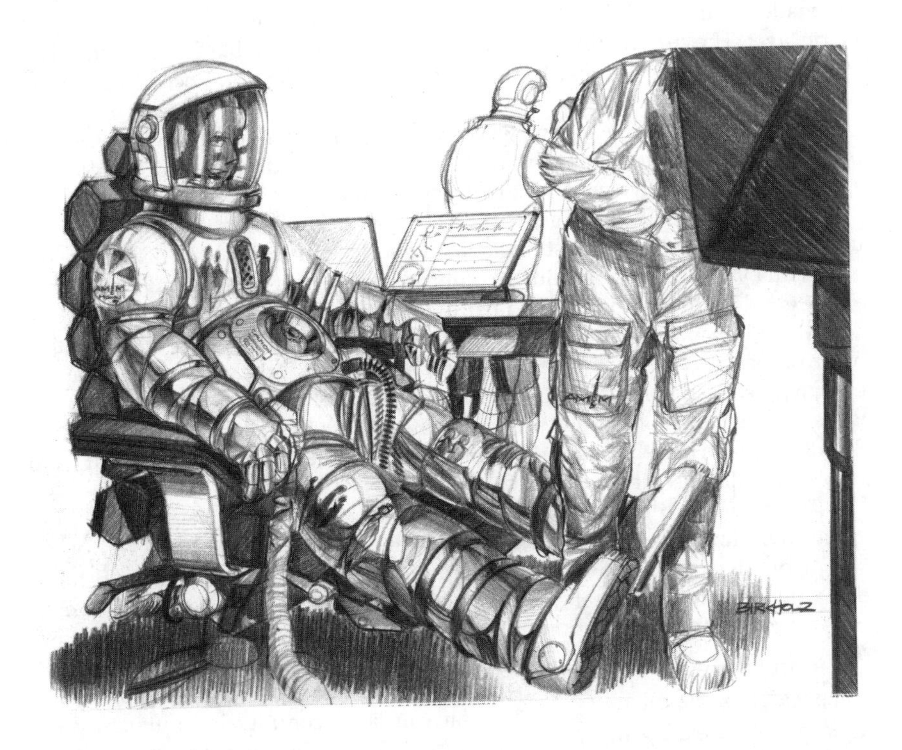

I was on my way to the office of Charlie Lyons, head of the crew systems group, when something caught my eye in one of the rooms I passed. It looked like a spacesuit. I stopped, wheeled around, and poked my head in the door. I recognized the three test pilots, although I had only met the senior of the group, Jake Hill; the other two had just recently come onboard. There was a fourth person whom I didn't recognize, and in their midst was a spacesuit. A beautiful spacesuit. Jake saw me and said, "Come on in—you'll want to see this!"

I didn't hesitate a second. Jake introduced me first to the fourth person, who turned out to be Bob MacNeely, the man in charge of AM&M's spacesuit program. Bob had spent 17 years working at Nuytco Research in Vancouver, British Columbia, Canada, on high-tech deep-sea diving suits, which, as I was to learn, had a lot in common with the new spacesuit. I was also introduced to the two new test pilots, Erica Phillips and Boris Feodorov. With that, we all turned our attention to the suit.

This was the first AM&M suit to be fully assembled, and Bob had called the pilots to take a look. It was smooth, shiny, and sleek—very different from the bulky, snowmobile-suit-like NASA spacesuits. I gently tapped the hard outer surface. "This is quite a bit different from the NASA and Russian suits that I've seen," I said. "Is there a fabric cover that goes over the outside—like those other suits?"

"No," Bob replied. "This is it. It's a hard suit—the NASA and Russian suits are made up of multiple soft layers of various undergarments, with an outer fabric layer. The current NASA model does have a so-called hard upper torso, but most of it is still very similar to the earlier soft suits. We don't need to put anything over this plastic exterior."

"It's plastic?" I asked, somewhat surprised. I had figured that it was aluminum, or composite, or maybe fiberglass. "What kind of plastic? Could you spare some time to give me a rundown on the suit?"

Jake said, "We've got to get to a meeting anyway. Thanks, Bob, for showing us the suit." The other two test pilots thanked Bob, too, and said their good-byes.

As they departed, Bob said, "Alright, come on over to my office. Now that we've got the suit together, I could use a break. Let's grab some coffee first."

Mugs in hand, we entered Bob's cluttered office. There were pieces of suits, other unrecognizable hardware items, and stacks of reports and technical papers everywhere. I had to remove a helmet from a chair before I could sit down. On the walls were pictures of various pressure suits and diving suits, from Wiley Post's Buck Rogers-ish high-altitude suit from the 1930s to sleek undersea suits that were very similar to the suit I had just seen.

"There are two things that drive spacesuit design," Bob began. "One, the space environment, and two, the spaceship cabin environment."

"How does the cabin environment affect the suit," I asked. "I would think that would be pretty well set—air for the crew to breathe, of course, much like on Earth—and then you put on the suit when you need to go outside."

"True, but as on Earth, your breathing air can be at considerably different pressures, depending on whether you're at sea-level, or on top of a mountain. The internal pressure and the composition of the cabin atmosphere have a big impact on the design of a space vehicle, and on the design of a spacesuit. Over the years, several combinations have been used." Bob took a swig of his coffee, and said, "Let me start with some history, eh?"

"The internal pressure and the composition of the cabin atmosphere have a big impact on the design of a space vehicle, and on the design of a spacesuit."

MANNED flight systems have always been part of the space mystique. After all, the question of how you keep people alive and comfortable in the hostile and unforgiving space environment is fundamentally intriguing. NASA had always played up the technical challenges. But, as information about the Russian space program came out in the 1980s, especially during the American experience on *Mir*, it became apparent that manned space systems could be awfully unsophisticated and still work. For the Apollo spacecraft, NASA had used liquefied gases to supply a breathable atmosphere for the astronauts, and this practice continued to the present day on the space shuttle. Russian spacecraft, on the other hand, had used superoxide compounds that reacted with the moisture and carbon dioxide breathed out by the cosmonauts to automatically produce oxygen for replenishment of the cabin atmosphere. These systems were the height of simplicity; the only maintenance needed was replacement of expended chemical canisters with fresh ones. There was one downside, though, as demonstrated on the unfortunate flight of *Soyuz* XII when a vent accidentally opened during reentry and the cosmonauts died of asphyxiation. There was no stored gas to replace what was lost. In the American capsules, enough excess gas was always available so that such a leak would not prove fatal.

On the other hand, the Americans had also made some tragic blunders in the area of life support. The first fatal accident in the NASA program was the Apollo I fire that killed three astronauts during a ground test; an electrical fire in the pure oxygen atmosphere quickly burned out of control. Earlier, in the Mercury program, a pure nitrogen atmosphere had been used to pressurize the cabin, with the astronaut's spacesuit providing oxygen for breathing. This system had been used on the X-15 as well. During one test in the Mercury program, though, a problem developed that allowed the inert nitrogen from the cabin to leak into the spacesuit air recirculation loop, which, at that time, was a simple rebreather circuit. In the rebreather circuit the oxygen in the suit was scrubbed to remove carbon dioxide, with new oxygen being added to make up for that which was consumed by the astronaut. However, during this test, nitrogen leaking into the circuit slowly began to crowd out the oxygen. The system worked by monitoring the gas pressure in the system; as the oxygen was metabolized, converted to carbon dioxide and removed, the pressure would decrease; more oxygen was automatically added to keep the pressure constant. Because of the nitrogen leak, though, the pressure stayed high enough to prevent the replacement oxygen from being added; if that went on too long, it would have been fatal. At the time, there was no instrument available for use in a spacecraft that could provide a real-time readout of the gas composition. The solution was to replace the nitrogen atmosphere in the cabin with pure oxygen for ground tests. Even at the time, it was a notably dangerous decision, which perhaps could be forgiven in light of the exigencies of the Cold War context of the space program. However, that same design philosophy was perpetuated forward, with little attempt to address the underlying hazard associated with a pure oxygen atmosphere in any closed human environment, much less a prototype spacecraft.

After the Apollo I fire, NASA switched to a mixture of 60% nitrogen and 40% oxygen on the ground. The addition of nitrogen reduced flammability significantly, and combined with scrupulous elimination of combustible materials, substantially reduced the fire hazard in American spacecraft. For in-flight use, however, they continued to use pure oxygen, although at a low pressure of about 5 psi; at that level, it had only about the same capacity to support combustion as normal air.

"Why did they use the lower pressure oxygen—why not just use air at sea-level atmospheric pressure?" I asked.

"That," replied Bob, "is the tie-in between the internal environment and spacesuit design. The early spacesuits had to be operated at relatively low internal pressure; otherwise, they would stiffen up like a balloon—the arms and legs would want to stick straight out—and that makes it hard for the astronaut to move, eh?"

"So why not use sea-level pressure in the cabin, and low pressure in the suit?"

"Well, ever heard of the bends?" Bob asked.

"Yes, that's something that divers get if they come up from the deep too quickly—oh. I think I see where you're going with this," I said.

"You're on the right track. Going from a higher-pressure cabin to a lower-pressure suit is analogous to a diver coming up from way below sea level to the surface. The higher pressure at which he has to breathe underwater means that nitrogen at high pressure will dissolve into bodily tissues and fluids. Since nitrogen is not used up by the diver's body, it accumulates—until he comes up. Then, at lower ambient pressure, that nitrogen will tend to come out of solution and form the bubbles that cause the discomfort, pain, and internal problems that constitute the bends."

"So if you have an oxygen-nitrogen cabin atmosphere, you will have some dissolved nitrogen in your body. And if you climb into a low-pressure suit—and the NASA suits operate at about 4.3 psi—the nitrogen can bubble out. So NASA kept the atmospheric pressure in the Apollo capsules and Skylab as low as possible, to minimize the pressure differential with the suits." Bob went on. "The Russians have always used sea-level atmospheric conditions in their spacecraft, and in their spacesuits a lower pressure, although still considerably higher than the U.S. models. They also use an oxygen prebreathing protocol prior to an EVA [extravehicular activity], to eliminate the nitrogen from the cosmonauts' tissues to help prevent the bends. This higher suit pressure system did restrict their ability to perform extravehicular activities. Actually, the first Russian spacewalk nearly proved fatal for cosmonaut Leonov. The combination of a stiff, high-pressure spacesuit and the need to use a small, deployable airlock because they didn't carry enough air to repressurize the entire *Soyuz* capsule made it almost impossible for him to get back inside. They got better at it after that."

"For the space shuttle, NASA finally adopted sea-level cabin pressure, but they weren't able to design a spacesuit for sea-level pressure that was not excessively fatiguing in a vacuum. In a high-pressure soft suit, you have to fight against the suit every time you move. For instance, if you want to bend your arm, you can only do so by reducing the volume of the suit, which means that you have to work against pressure in the spacesuit. The higher the pressure, the more effort it takes. Doing any physical work while wearing one of these suits will tire you out very quickly. As a compromise, they too went to an extensive oxygen prebreathe; they also go to an intermediate lower cabin pressure during the period just before an EVA, in order to get by with a low-pressure suit. Both of these are time consuming, and the reduced pressure makes it harder to keep the shuttle avionics and electronics systems cooled properly. However, they're still doing the same things on the ISS [International Space Station]."

"Doing any physical work while wearing one of these suits will tire you out very quickly."

"But you're not—on the DH-1, right?" I asked. As I had been learning, AM&M as a whole tended to take a close, hard look at any standard space system design or operational practice, and they never shied away from taking a different approach when it made sense.

"No way," Bob responded quickly. "For routine, commercial operations we need to eliminate the possibility of the bends and get rid of the prebreathe.

Our crews and customers will also want much better mobility and protection than the NASA suits provide. Besides, those things are too darned expensive, eh?"

BOB proceeded to tell me about some of the more interesting alternatives to the standard spacesuits that NASA and the Russians had been using for the past 40 years, and that lately even the Chinese had copied. John Forsyth, of course, was aware of the problems with the current suits as well as the alternative approaches, and he had specifically asked Tom to have the crew systems people go back and reexamine the earlier work. Essentially, there were two really innovative spacesuit concepts that seemed to have potential. One was known as the Webb suit, after its inventor, Paul Webb. It was so radical that it never received serious consideration from NASA, or anyone else, outside of science fiction. On the other hand, it was the simplest, lowest cost suit imaginable. It was based on the concept that, instead of surrounding the body with a miniature spaceship, you supply a suit of clothing suitable for the space environment.

When venturing out at −100° at the South Pole, station personnel do not don miniaturized habitation cubicles that they carry with them. They wear instead foam-lined boots and thick down jackets; the dominant problem is the cold. In space, the dominant problem is lack of pressure. If it were only a lack of breathable atmosphere, something akin to SCUBA gear would be adequate. Webb's insight led him to design a garment that put pressure on the skin to balance the body's internal pressure. A helmet, covering the head and neck, was sealed to the skin with a rubber gasket, and a rather sophisticated cup was used to cover the genitals. The rest of the body was encased in what amounted to several layers of very tight hosiery. The system was all the more remarkable when you considered that the garment was not even airtight. Sweat from the skin could wick out through the cloth, evaporate in the vacuum of space, and provide the necessary cooling for the wearer. The harder you worked, the more sweat you would produce, and the more you would be cooled. If additional protection from solar radiation were needed, an outer garment could be worn over the suit without interfering with its cooling capabilities.

The major downside of the Webb suit was that, even with the cup, it tended to grab male wearers uncomfortably in the crotch. It was also only practical with low-pressure, high-oxygen-content breathing air, and so would require the extensive prebreathing of other low-pressure suits, or the use of an oxygen-helium atmosphere within spacecraft, which makes people sound funny when they talk. Still, cost estimates at the time (the late 1960s) came out to less than $10,000 per suit, or some $60,000 in today's dollars, which made this approach worth investigating again, especially if a space tourism market ever did evolve.

The second intriguing approach was the Vykukal spacesuit. Hubert "Vic" Vykukal, working at NASA Ames Research Center in the 1960s, had developed a hard suit, without flexible joints, for use on Apollo 20 and beyond. The early Apollo suits, like all other soft suits, suffered from the problem mentioned before—you had to work against the internal pressure every time you moved, which made them less than ideal for lunar exploration. In addition, the suits were composed of multiple layers—a pressure garment, an external woven Teflon® protective garment, and an inner water-cooled suit—and they were not particularly easy to don or doff. More or less handmade one by one, these suits ended up costing millions of dollars each. Bob told me that someone had once figured out that the NASA suits actually cost over $50 million apiece, if you divide the total spacesuit program expenditures by the number of suits actually built.

The Vykukal hard suit was based on the technology pioneered for deep diving suits by Phil Nuytten, Bob's former boss and founder of Nuytco Research. These suits allow divers to work underwater at atmospheric pressure inside their suits when the pressure of the surrounding water is as high as 30 atmospheres—with no risk of the bends. With such incredibly high external pressures, you obviously can't have a suit that could be allowed to change in volume; otherwise, the outside pressure would crush it. So it had to be a hard suit, and Nuytten had to devise some other way for the wearer to move his limbs. He came up with a series of rotating joints arranged so as to allow both rotation and bending motion. The rotating joints are angled with respect to a limb, with two halves each comprising a thick wedge section and a thin section. When the wearer bends an arm, the joints rotate so that the thin sections come together, allowing the arm of the suit to assume a bent position. Similar joints are commonly used in stovepipes and ductwork, where joints can be created at virtually any angle merely by selectively rotating the several segments. Anyone who has seen the movies of men doing pushups in hard suits cannot fail to be impressed by their flexibility, as well as by the strange kinematics associated with rotating the joints to allow bending motion.

For a hard spacesuit, of course, a 14-psi pressure differential is nothing compared to the 450-psi differential that the deep diving suits have to withstand. Vykukal's follow-on Apollo hard suits had been constructed of fiberglass segments. Of course, Apollo was terminated with Apollo 17, and so the Vykukal suit was never used. During the early 1980s, he built a prototype aluminum suit that could provide its wearer with radiation shielding in geosynchronous orbit. The latest NASA suits incorporate joints with some rotational capability, as well as a hard upper torso, but by and large Vykukal's suit has been ignored. The conventional suits were well established in the crew systems' organizations within NASA, and there was little incentive to adopt a new approach.

The suit being developed by Bob and his team at AM&M was based on Vykukal's concept, using an internal pressure of 10 psi. This is high enough to eliminate the possibility of the bends in going from the crew cabin at 14.7 psi into the suit. But the fabrication technique was notably different. Bob had come up with a way to use a modified blow-molding technique for the plastic suit. Blow molding is used for a wide variety of hollow plastic consumer products, such as bottles and trash cans. The plastic Bob had selected was similar to that used in the familiar 2-liter soda bottles. Those bottles are actually composed of a tough plastic, PET (polyethylene terephthalate), interspersed with layers of PVA (polyvinyl alcohol) that is resistant to migration of the CO_2 used to carbonate the beverages. The two plastics, which are immiscible, are blended together. During the blow-molding process, the PET and PVA separate into layers called lamellae. These 2-liter bottles easily withstand more than a 32-psi pressure differential and weigh less than 1% of the weight of the soda that they contain. For the spacesuits, where the cost of the plastic is relatively insignificant, a blend with much better material properties was selected, which provided higher strength and greater resistance to punctures and also to UV light.

The blow-molding operation was supplemented by an injection-molding process that formed the elastomeric seals on the segment joints. For hollow parts, a blow mold is usually quite a bit cheaper than an injection mold because the pressures in the mold are considerable lower. Fortunately, the injection molding of the seals took place in a very limited area of the mold, and when it was all said and done the molds cost an average of about $50,000 apiece. The design of the AM&M suit was modular, so that by combining different plastic pieces with the aluminum rings that held the joints together, spacesuits could be assembled to fit anyone between 5 feet and 6 feet 2 inches tall. In all, about 70 different pieces had to be molded. The total cost of tooling came to about $5 million—not much more than the cost of a single NASA suit. Including the other parts and subsystems, plus the design work—most of which had been done in-house with the assistance of some former Ames people and some of Bob's colleagues from the diving industry—the total cost per suit was coming in under $50,000, even at relatively low production volumes.

The plastic blend that Bob had selected produced crystal-clear plastic pieces. Using a chemical process much like the one amateur telescope makers had employed to coat their mirrors before aluminum vapor deposition was widely available, the interiors were coated with silver. A durable, scratch-resistant paint was then applied over the silver to protect it and keep it from tarnishing. The result was the stunning, silvery suit that had caught my eye as I walked down the hall.

The shell of the suit was thus similar to the standard heat-rejection panels used on satellites. The silver reflects the incident solar radiation, which is the main source of heating, but the thin transparent plastic wall over the silver also acts as a radiator in the deep infrared and therefore functions to reject body heat. Altogether, in Earth orbit, the AM&M spacesuit could reject 400 Btu per hour (about 100 watts) without an active cooling system. Of course, if the astronaut is physically active at more than a minimal level, the heat-rejection requirement can increase by a factor of four or more. For additional heat rejection, a cooling system similar to that used on the Apollo suits had been included.

THERE are two basic ways to cool spacesuits. One is to force dry air through the suit to evaporate the astronaut's sweat and remove the excess heat as the latent heat of vaporization of water. Air is blown down at the feet and then allowed to flow upward over the astronaut's body before being dumped overboard. This approach, which was used for the Gemini program, is not the most mass-efficient solution because the required airflow is much more than is necessary for breathing. It is, however, the simplest solution.

The second, more efficient approach uses a water evaporator exposed to the vacuum of space. Such coolers were used in the Gemini capsule for cabin air, with a bimetal valve that opened a vacuum port to the evaporator, depending on the temperature of the water circulating in the coolant loop. The problem with this approach was that the valve could potentially get stuck in the open or closed position, resulting either in excess cooling or total cooling failure. For Apollo, a simple porous nickel-plate system had been developed. Water in the coolant loop was circulated through an evaporator that consisted of a porous nickel plate exposed to the vacuum. The water then diffused through the plate towards the vacuum where it would evaporate and then freeze, thus sealing the pores. As warmer water from the loop heated the nickel plate, the ice would sublime—like dry ice going directly from solid to vapor—and absorb heat from the nickel plate, which absorbed heat from the circulating water. As the pores of the nickel plate opened up, more water would flow into the pores until they began to freeze up again. Thus the porous nickel plate provided a very simple, self-regulating evaporative cooler for the Apollo suits.

The AM&M suits were further simplified by limiting the water-cooled portion to a vest that was worn around the central chest cavity, so that the astronaut's breathing provided the pumping action to circulate the water through a system of one-way valves in the coolant loop. The cooling water then circulated through a porous nickel plate on the chest. The harder the astronaut worked and the faster he breathed, the faster the water was circulated through the cooling plate. Bob was not sure why nickel plate had been selected for the Apollo capsule, but he suspected that it was a spin-off from the atomic-bomb development program. That had required endless miles of porous nickel tubes to enrich uranium in the gas diffusion process. Undoubtedly, this had created a reservoir of experience in the accurate design and fabrication of nickel plate of a desired porosity. Extra comfort was provided by increasing airflow somewhat so that sweat would be removed from the feet and extremities. However, the entire cooling load could be handled by the chest pack evaporator.

For simplicity, the breathing air system was designed to operate like SCUBA gear, using compressed air that was circulated through the suit and then dumped. For longer-duration extravehicular activity, a backpack containing recirculating fans, 12 pounds of additional cooling water, 3 pounds of oxygen, and chemical cartridges to scrub out the excess CO_2 was under development. A man doing moderate work consumes about 2 pounds of oxygen per day and produces an amount of CO_2 that can be absorbed by about 1½ pounds of lithium hydroxide. Each pound of water will supply 1000 Btu of cooling capacity to supplement the basic cooling capability of the suit. The backpack was expected to weigh less than 25 pounds, and would keep an astronaut comfortable for 8 hours, and keep him alive—if less comfortable—for up to 24 hours.

DURING launch, and when wearing the spacesuit within the vehicle, air would be provided through an umbilical that tapped into the high-pressure air supply for the crew/payload bay and the cold-gas reaction control system. The DH-1 payload bay would be pressurized at 20 psi from launch until the vehicle reached an altitude of about 15,000 feet, when it would be lowered to sea-level pressure and maintained at that level until arrival on orbit. For payload deployment, the entire bay would be vented to space and the payload manually ejected through the payload bay door. After deployment, the door would be closed, and the crew/payload bay would be repressurized. Its 1100-cubic foot volume would require about 110 pounds of air. The pilot, using a simple open-loop SCUBA-type air supply, where only 5% of the gas is actually utilized (most of it being dumped overboard), would need an average 5 pounds of air per hour, or another 120 pounds for a 24-hour on-orbit stay. A separate 15-minute supply of high-pressure air was mounted on the pilot's seat to supply air in the event of an ejection.

Use of high-pressure gas was integral to the simplicity and reliability of both the life support and the reaction control systems. In fact, both systems were supplied from the same storage tanks. The spacecraft actually had two reaction control systems, one consisting of a set of small, cold-gas jets and another using hydrogen-peroxide and propane thrusters. The cold-gas system, using air, produced about 70 seconds *Isp*; the hydrogen-peroxide/propane system operated at about 200 seconds *Isp*, providing greater thrust to ensure control over the vehicle during reentry. At this point, Bob digressed to explain the technology that made high-pressure gas systems feasible without a large weight penalty.

The weight of a container for high-pressure gas is dependent on the strength of the material of which it is constructed. Of course, the strongest containers are those employing engineered "superfibers," such as graphite and Kevlar®. However, these fibers (within a suitable resin matrix) cannot be used by themselves because high-pressure gas tends to leak through the composite walls. Extremely lightweight tanks can be fabricated of aluminum that is overwrapped with high-strength fibers. The problem with this approach, though, is that in order to load the fibers—in other words, to make use of their strength—the consequent elastic stretching of the fibers results in yielding of the aluminum liner. Thus, each time the gas bottle is charged and discharged the internal liner yields. This results in a relatively short fatigue life. It is workable for one-time applications. Burt Rutan had used this technique in the oxygen system of the Voyager for its around-the-world flight, but those tanks were never used again.

The solution is to use a higher-strength material such as stainless steel or titanium, which yields the first time the bottle is pressurized, and then, when it is emptied, goes into compression. So instead of just using the metal's positive elastic range, you can use the entire range from negative to positive. Such gas bottles, made of high-strength stainless steel and cryo-yielded during the first loading, can be reused for many cycles, with a figure of merit of one million cubic inches-psi per pound. One problem, however, is that only about half the strength of the bottle comes from the composite overwrap. If the liner is made too thin, it will buckle when the bottle is empty. Lockheed Martin had improved on this approach by bonding the metal liner to the composite overwrap with sufficient strength so that the thin-walled liner doesn't buckle when the bottle is empty. Here AM&M had adapted the technology used for inspecting the main propellant tanks. A simple electronic circuit generates an ultrasonic pulse in the wall of the liner during the low-pressure part of the cycle, which is used to detect any buckling that has occurred. With this simple and reliable technique, such gas bottles can be safely used in a manned system. Figures of merit of up to two million can be achieved. In practice, this works out to about 0.2 pounds of bottle for each pound of high-pressure air. The orbital stage has a single high-pressure gas storage tank capable of holding 5000 cubic feet of air at a pressure of 10,000 psi that weighs only 100 pounds. This includes the high-pressure regulator that supplies 2000-psi gas to the cold-gas RCS jets, and the low-pressure regulators that supply 10 or 14.7 psi to the spacesuits and crew cabin/payload bay, respectively.

The life support system was being designed for a nominal 48-hour mission, with two cabin volumes' worth of air being available to allow opening the payload bay door by depressurizing the cabin and then repressurizing it after payload deployment. An additional one volume's worth would be stored in reserve. For a nominal volume of 1100 cubic feet, the cabin air would contain 230 cubic feet of oxygen. The air circulation system included rechargeable dehumidifying cartridges and lithium-hydroxide canisters for CO_2 removal, which served to keep the air within the cabin quite comfortable. The dehumidifiers could be recharged by exposing them to the vacuum and heating them gently with built-in electrical heaters. With the CO_2 removal capability and an average usage rate of 1 cubic foot of oxygen per hour per person, approximately one-and-a-half days' worth of oxygen was available for a crew of two, without adding any additional air from the high-pressure supply; roughly one-third of the oxygen could be used up before it became noticeable. In a normal office building, the air exhausted by the ventilation system is about 1/2 of 1% carbon dioxide. Breathing distress does not really become apparent until somewhere between 3 and 5%. Using 2% as a reasonable set point and with a single pilot onboard, the DH-1's air purification system could even be shut down for up to 5 hours without concern.

In a serious emergency, with the use of the lithium-hydroxide canisters, three days' worth of oxygen could be consumed from each volume of cabin air, albeit at some discomfort and with reduced mental alertness for the pilot. Enough lithium-hydroxide canisters would be onboard for at least eight days on orbit using the three available cabin air exchanges. The eight-day capability was based on preliminary logistics planning that indicated that a rescue mission could be mounted within four days' time, at a confidence level of 80 to 90%, from any base located at a latitude lower that the stranded vehicle's orbital inclination. Undoubtedly, at some point in the future when enough vehicles were in service, some agency—perhaps the U.S. Coast Guard or Air Force—would keep at least one vehicle always on alert at an equatorial site, which would reduce the response time to hours instead of days.

At sea-level pressure, 1100 cubic feet of air weighs approximately 110 pounds. For very weight-critical missions, the vehicle could be operated with a helium-oxygen atmosphere, which would save approximately 60 pounds. With three cabin volumes' worth of air being carried for a typical mission during which the payload bay was evacuated and repressurized, over 170 pounds could be saved. In the long term, when most flights would involve docking with a space station, air could be pumped over to the station; even at $100 dollars per pound, 110 pounds of air would represent $11,000 worth of cargo on orbit.

... even at $100 dollars per pound, 110 pounds of air would represent $11,000 worth of cargo on orbit.

FOR food and drink, there would be collapsible water bottles and drink mixes, candy bars and granola bars, fruit, and sandwiches. A simple table with a porous surface could be placed over the air return intake, so that during a meal food particles would not spread throughout the interior of the spacecraft. With a nominal on-orbit flight time of one day, though, food was not a major concern.

However, for future contingencies, something else might be needed. Forsyth had always been interested in the sledging rations used by the early Antarctic explorers. Weight had been very critical, and so the rations had consisted mostly of fat and dried meat. Something like that would be needed for emergency rations in space. Air and water could be recycled relatively simply, but in the near term food was going to have to be carried along. Fats or oils, of course, are the most weight-efficient foods; one pound of oil contains about 4000 calories, whereas a pound of starch or sugar contains only about 1500 calories. Unfortunately, rations based largely on fat tend to lead to that dreaded occupational disease of the arctic and Antarctic explorers—hemorrhoids. Thus the crew systems group was working on developing a ration with a high oil content, but containing sufficient fiber, minerals, and other additives to meet the complete nutritional needs of an astronaut for long-term health. The goal was a 2200-calorie daily ration that was at least palatable and weighed no more than 12 ounces, including packaging. These lightweight rations could be used for ballast to balance payloads and thus be transported to orbit very cheaply.

Unfortunately, rations based largely on fat tend to lead to that dreaded occupational disease of the arctic and Antarctic explorers—hemorrhoids.

A manned system, of course, requires a bathroom. I was aware of all of the problems that NASA had faced with their shuttle toilet, and so I just had to ask Bob that perennial question. "How will DH-1 crews and passengers go to the bathroom in space?"

"The short answer is," Bob replied, "with some difficulty. Toilet functions really do benefit from having a gravity field to work in, eh? That makes it so much easier to get the stuff to go where you want it to go, and to keep it from going where you *don't* want it to go."

He gave me a brief review of the history of waste elimination in space vehicles, and it wasn't pretty. From Alan Shepherd's need to urinate into his spacesuit during an extended hold in the countdown for his first space flight, to Apollo's plastic bags that had to be fastened to the buttocks with adhesive for defecation, to the space shuttle's infamous toilet—it all sounded rather distressing for a would-be space tourist.

Bob told me how NASA's original design for the shuttle had involved the "shit hitting the fan," literally. The shuttle toilet handled solid waste by shredding it and throwing it against the wall of the vessel, where it was collected and dehydrated. Unfortunately, the next time an astronaut used the toilet and the shit hit the fan, it sent some of the freeze-dried material spewing out of the toilet. With the high humidity levels in the shuttle cabin, this resulted in the fecal material plating out all over the shuttle interior. Not nice! They finally modified the system to include a highly absorbent diaper-like material through which air was drawn to collect the waste, which vastly improved the situation.

Urination was always easier, with relief tubes and urine bags providing a simple solution. After the suborbital Mercury flights, and Shepherd's unpleasant experience, spacesuits also incorporated this feature. For later suits, including the AM&M model, the tube-and-bag system was replaced with a superabsorbent diaper much like those used by adults who have to deal with incontinency. This served to impose a certain egalitarianism between male and female astronauts, but more practically, it eliminated the possibility of a broken urine bag, and the consequent necessity of trying to make your way back to your ship with urine all over the face plate of your helmet.

For the DH-1, the toilet was simplicity itself. It too used ultra-absorbent diaper-like liners, backed with activated charcoal to remove odors and was connected to the vacuum side of the air circulation system to move the waste in the proper direction. After each use, the liner was rolled up and sealed in a bag, like a dirty baby diaper. No doubt a lighter-weight solution would be required in the future for long-duration flights, but for now this system was simple and effective. It was cheap, too; development costs were considerably less than $100,000.

"So you see, we've come a long way since the early days," Bob concluded. "Not as comfortable as your own bathroom at home, but if you want to go *to* space—you can *go* in space without too much unpleasantness."

Not completely convinced, I said, "I guess that doesn't sound too bad."

CHAPTER 17
Markets, Philosophy, Techniques, and Approaches

AFTER my flights in the simulators and my discussion with Bob MacNeely about the spacesuits and crew systems, I spent the next two weeks with Forsyth and Tom on a literal round-the-world tour as they paid courtesy calls on an eclectic group of government officials, aerospace executives, industrialists, air forces, and space enthusiast groups. This was the first phase of Forsyth's marketing campaign. At this point, without any flying hardware, AM&M's credibility was naturally low, and Forsyth wasn't expecting anyone to be signing up to buy vehicles any time soon. However, he needed to plant the seeds. They had an interesting presentation, an engaging story, and a colorful set of PowerPoint slides. As long as it didn't cost them any money, air force generals, government bureaucrats, members of the legislature and executive branches, industrialists, and aerospace managers were willing to listen. Forsyth's intent was to build a foundation of understanding for AM&M's technical approach, so that when successes were achieved the engineering behind them would be credible. The key to gaining that credibility was to be understated, conservative, and always do what you said you were going to do.

Forsyth's plan was to create an overall favorable public awareness of the company and the DH-1, while gradually developing a consensus that space was going to be the next important area of economic growth. Then he wanted everybody to feel that they had to get on the bandwagon if they didn't want to be left behind. It was this sort of thinking that had made the investment capital available for everything from railroads to personal computers, disk drives, and the Internet, and Forsyth was counting on it to provide the money to underwrite the purchase of the vehicles by their future customers. Owning a DH-1 was going to be seen as key to being a player in an emerging industry, which, for a number of years at least, was going to be mostly sizzle and very little steak. Forsyth anticipated that the cycle would be much like that in other new industries: high expectations and excitement would make lots of capital available, which would lead to a rapid expansion of capabilities, followed by a consolidation that would set the stage for long-term growth. Profitability for the industry as a whole might or might not show up for decades. But emergent industries can grow rapidly and become economically significant even if many or most of the players never achieve profitability. On the contrary, huge sums can be lost—as so many had learned in broadband.

Owning a DH-1 was going to be seen as key to being a player in an emerging industry ...

In making these presentations, though, Forsyth knew he was walking a tightrope. The success of their venture would turn the whole aerospace industry upside down, resulting in the cancellation of many existing programs and the loss of billions of dollars of government contracts. If people started to believe that what was described was really going to come to pass, some of them would be motivated to take action—political, economic, legal, or simply competitive—that would be detrimental to the interests of AM&M. Forsyth had once been warned by a man with considerable marketing experience in the aerospace industry that the big boys play rough, given the billions of dollars that are at stake. Of course, Forsyth firmly believed that if the AM&M program were successful, the new initiatives and new money spent in the space business would dwarf that which was lost. However, there were certainly no guarantees that today's major players would be tomorrow's major players. In Forsyth's experience, people tend to be more concerned about what you are taking away than what you might be giving them.

So although the presentation was exciting, and laid the groundwork for an appreciation of what John Forsyth hoped would be an unending series of successes, it was calculated to de-emphasize the negative impacts on existing systems and institutions, and the presentation was not so detailed that "older and wiser heads" could brush off the whole concept if they did not want to believe in it. By and large, aerospace executives and government official are a conservative lot. They were unlikely to believe that one more brand-new launch-vehicle company in the long line that had come parading before Wall Street, industry, and the government agencies, pedaling one more design, was going to amount to much. Forsyth was counting on the myopic vision of those who might be hostile to the success of the DH-1 to keep them from comprehending the likelihood of that success and the consequences that would flow from it. They would thus not be motivated to take any vigorous counteractions, at least until success was so far along that fighting it wouldn't be worth the effort.

On this same tour, while developing a controlled level of expectation and interest within government agencies, air forces, and aerospace companies, Forsyth and Tom were also working hard to form beneficial industrial alliances around the world. When the time came to sell the vehicle to the Japanese, or the Chinese, or the Indians, or the French, or the British, they would find that their country already had a substantial investment in the vehicle and a stake in its success.

Forsyth was aided in accomplishing his goal of not awakening the sleeping giants by the patently ridiculous nature of the AM&M vehicle—at least from the conventional perspective. Its payload capability was puny, it couldn't haul any major existing payload to orbit, and it just wasn't high tech enough. When compared with even a decades-old system like the shuttle, a high-performance hypersonic airplane with many advanced and sophisticated components and systems, the DH-1 looked small, simplistic, and even primitive.

NASA, with over $2 billion appropriated, was just about to enter the preliminary design phase for a two-stage winged vehicle, the first stage of which was to fly to Mach 6 before zooming out of the atmosphere to launch a second winged orbital stage. It was brimming with new technologies, and NASA was promising lower costs, greater reliability, and enhanced safety. Every major wind tunnel in the country, from subsonic to hypersonic, and all of the big (and many small) aerospace companies and research institutes were working on the program. Serious people were just not spending a lot of time worrying about whether AM&M was going to build a successful reusable launch vehicle that would completely revolutionize the way things were done in space.

It was Forsyth's personal opinion that this latest round of NASA contracts and development projects would be like the previous three or four—it was hard to keep track of them—next-generation launch vehicle fiascoes. The most egregious of these was undoubtedly the National Aerospace Plane, or NASP, program of the mid-1980s. NASP had consumed $3 billion without producing so much as a decent mock-up vehicle. The game generally seemed to play out as follows: do some technology development, run into a few road blocks, decide not to waste any more money on that program, and then retreat to consider the options. Forsyth was hoping that during one of the retreat phases, the DH-1 would make its debut and put an end to the quest for exotic, futuristic second-, third-, or fourth-generation reusable launch vehicles. Tom figured that at least this latest NASA effort should provide some interesting technological developments, which AM&M would be free to cherry-pick to the extent that they fit within its game plan.

We were holed up in an airport bar in Frankfurt, waiting for our second delayed flight of the evening. Cocktails in hand, we were hashing over the latest *Aviation Week* cover story about the NASA vehicle development program. It occurred to me that although it was clear to everyone at AM&M that this program was never going to provide a practical, low-cost reusable launch vehicle, what was to keep NASA from doing what they had done in the past—tell people a rosy story about how wonderful and cheap it was going to be and how they should wait and buy or fly a NASA-developed launch vehicle, which would be available real soon now.

"Well," John Forsyth said, "they probably will try that again. But NASA still has some problems with credibility. While there are not too many people that doubt that NASA can do some pretty amazing technical feats, the cost overruns in the International Space Station program are still talked about. Another thing in our favor—and which should be pretty obvious to anyone who might actually be in the market to buy a space launch vehicle—is the difficulty that NASA or even one of its contractors would have in ever selling anything."

Chris Tombal, an engineer traveling with us on this leg of the trip, had just come back to inform us of the latest delay, and so we ordered another round. After our drinks arrived, I asked, "Ok, I know NASA doesn't sell vehicles. But why can't the contractors sell them? They are, after all, private companies."

Tom spoke up. "Government contracts are based on paying the *cost* of something. Therein lies the rub. That was one of the first things I learned about the aerospace industry on my first job. I asked one of the old hands to explain the aerospace procurement system, which seemed to involve way too many people for the work to be done, and costs that were higher than they ought to be. This guy took me aside and told me about how, during World War II, some aircraft and defense companies had made large profits. That didn't sit too well with the public or the Congress, so the government decided that they were only going to pay what things cost, plus a reasonable—read, small—profit, and no more."

Tom went on to explain that, for the most part, the military procurement bureaucracies had stuck with that system. The problem was that the profit motive controls costs and so a system geared to controlling profits eliminates the most efficient means of controlling costs. If a company wanted to double their profits under a cost-plus system, they need only double the cost of what they were building. This was rather easy to do in the defense industry by simply not resisting the various specifications and regulations, set-asides, and overmanagement that government contracting tended to impose. Of course, when the government needed to get something important done in a hurry, they used organizations like the Skunk Works, which, operating in a classified environment, meant that the fact that somebody might be well paid or that the company made a decent profit could be kept from the public eye. Really amazing things were done relatively cheaply. But by the late 1990s, the Skunk Works-type operations had been largely absorbed into the mainstream, "business as usual" parts of their parent companies. So the government always paid what something cost, never what it was worth.

The CEO of Northrop had explained another major problem with this approach to Tom back in the 1980s when that company was trying to sell its F-20 lightweight fighter to both the U.S. Air Force (USAF) and allied air forces. If you develop a new product and try to sell it to the government, they want to pay exactly what each unit cost. But the first jet fighter off the line might cost $100 million, whereas the hundredth might cost only $15 million. The government was perfectly happy to start out paying $100 million and eventually end up paying $15 million per fighter. But of course, an overseas customer wasn't about to pay $100 million for an airplane that was worth $20 million.

Obviously, the first airplane wasn't worth $100 million; it just cost $100 million. But that is true for almost any new product. The first car off the assembly line costs a heck of a lot more than the hundred thousandth or the millionth car. This is one of the reasons why a really high-end sports car costs 10 or 20 times what a perfectly adequate family sedan does, simply because the numbers are so much smaller for the sports car. When Northrop wanted to sell the F-20 overseas, they could only ask what the plane was worth. If their costs were lower than the value of the plane, that meant they made a profit. But if the U.S. government were also buying the planes (and sales to the USAF were necessary to launch the new airplane), they couldn't—by law—be paying $100 million for an identical item that somebody else was buying for only $20 million. It's a classic Catch-22 situation. The system worked for mature production programs, where the cost was already at or below the value of the system, and so it could be sold at or above what the U.S. government was paying for the same aircraft, and hence provide a reasonable profit. But a new system couldn't be sold to foreigners or even to a different U.S. government agency for less than what the primary government customer was paying. No service secretary, no congressional committee would put up with that! The bottom line was that, even if NASA did build a new reusable launch vehicle, and even if they had good value, they were going to *cost* several times their value for quite some time—and actually forever, if no more than four or five were built. It just wasn't possible in the existing government procurement environment to sell such a vehicle at what it was worth, if that was substantially below what it cost.

Of course, it was beyond anyone's belief that NASA's first vehicle—or even its fifth or tenth vehicle—was going to cost less than its intrinsic value. And a technically sophisticated, two-stage, hypersonic winged vehicle was by its very nature going to require a substantial infrastructure and have a high unit cost, regardless of whether you built five or 100 of them. Even the Concorde supersonic airliners, which saw 30 years of commercial service but had cost some $70 billion in today's dollars for the development and production of 22 airplanes, never had a value greater than their cost. No, it wasn't likely or even believable that NASA and its contractors would ever be able to offer anything at a competitive price. Nor was there any economic advantage to be gained by placing a small, winged reentry body on an expendable booster, as the Japanese were poised to do, with the French, the Chinese, and even the Russians not too far behind. Placing a winged reentry body on an expendable just reduces its useful payload by a factor of four to seven, while giving you very limited manned access to space.

Even the Concorde supersonic airliners, which saw 30 years of commercial service but had cost some $70 billion in today's dollars for the development and production of 22 airplanes, never had a value greater than their cost.

Of course, Forsyth didn't expect the first DH-1 to cost less than customers would be willing to pay for it. That, he explained, was the reason behind the completely separate company that was being set up to sell them to the U.S. government. The government was allowed to buy catalog items at catalog prices; it was just that when something was built to government specifications or modified to government specifications that it came under the procurement laws. By maintaining an arms-length relationship between AM&M and the military/government supplier, AM&M could avoid entanglement with the federal acquisition regulations, or FARs. This second company would buy the vehicles from AM&M and then do to them whatever the government wanted done: modifications, special equipment, compliance with military specs, etc. On completion, they would resell them to the government, with all of the additional features and the additional paperwork necessary to comply with the FARs. This company would not be a subsidiary of AM&M; in fact, Forsyth was prepared to sell to vehicles any company that wanted to resell DH-1's to the federal government. There were lots of useful, even exciting things—like on-call orbital rescue, orbital debris removal, and geosynchronous traffic control—that would be legitimate government functions because they weren't likely to generate income streams sufficient to support themselves. But everyone would be happy if the government did them. Thus, he hoped to sell more than a few vehicles to the U.S. government. However, AM&M would get its market price—the same price at which it was selling to everyone else—and its profits. John Forsyth at least hoped that those profits would be well in excess of what the government considered reasonable.

Chapter 18
Rockets, Jets, and Soft Landings

AFTER recuperating from the road show with Forsyth and Tom, I took a flight out to Mojave, California, where Tom had a small team working on a jet-engine landing system for the first stage. Rocket-powered landing had been demonstrated by the DC-X, and later by the Japanese RVT, and was still the baseline for the prototype because it seemed to be the most mature and readily available technology. There were, however, some advantages to using jet engines instead, mainly related to a higher confidence in getting the first stage back to the launch site.

With rockets, the prototype stage in fact could only handle about a mile of unanticipated lateral offset as a result of winds during boost, coast to 350,000 feet, and descent. This effectively imposed rather stringent operational limitations, restricting launches to periods with relatively mild, steady winds. Depending on the season, the trajectory people were predicting that launch conditions would be satisfactory only about 20% of the time at Cape Canaveral and worldwide perhaps 30–40% of the time. Overall, the probability of missing the landing target remained at about 3–5% for the rocket system (even during that 20 or 40%), which would restrict the vehicle to large ranges or relatively unpopulated areas, and require a 3–5-mile-diameter landing area.

In contrast, with airbreathing engines for landing, simulations were showing that wind conditions would be suitable at most launch sites most of the time, with a correspondingly greater probability of a successful landing at the launch site. Forsyth was thus quite interested in the development of the jet (actually turbofan) landing system. At this point, however, he was more interested in getting to orbit rather than getting the last bit of efficiency and operational convenience from the vehicle; hence the decision to stick with rockets for the present. Although he was confident that he could sell vehicles with the rocket landing system, for long-term operations and the very high launch rates that would be necessary to drive the cost to orbit below $200 per pound, a system that could operate almost all the time was definitely going to have an advantage.

The primary difference between the two options was the time that the engines could be operated: the longer the duration, the more capability to fly to the desired landing point. Using rockets for landing, the DH-1 would be able to carry only enough propellant for about 30 seconds of thrust. That wasn't a lot of time or a lot of power with which to fly the vehicle back to the landing site, if winds during the descent carried it too far off the predicted track. The duration of powered flight is dependent on how much thrust you can produce per pound of propellant burned per second, which is another name for the *Isp* of the propellant/engine combination. The hydrogen-peroxide monopropellant engines had an *Isp* of about 160 seconds, and the restartable liquid oxygen-methane engines would have a sea level *Isp* of some 280 seconds. Airbreathing engines, however, could easily achieve 4000 seconds, and if specially designed engines with a suitable bypass ratio were developed you might get 10,000 or even 15,000 seconds. So for an equivalent propellant weight that with a hydrogen-peroxide rocket engine would give you 30 seconds of operating time, an airbreathing engine could offer 20 to 60 minutes, or more than 10 minutes with only one-half, or perhaps even down to as little as one-sixteenth, the fuel weight. That would allow the vehicle to fly to the landing point even if it were 10–20 miles off target. This would afford, it was estimated, the ability to operate in sustained winds up to 35 knots, with gusts up to 50 knots. That would make launches possible about 90% of the time at Cape Canaveral and 95% for most other sites.

The downside of the airbreathing engines was the lower thrust-to-weight ratio (T/W) and correspondingly heavier engines because you still needed the same maximum thrust. The jet engine group had determined that an engine with 8000 seconds *Isp* and a T/W of about 15:1 was a doable near-term goal, in view of cost and overall weight considerations. They had Pratt & Whitney working on a high-performance lift engine based on an existing engine core, but the time required before it would be available was still estimated to be three to four years—in part because funding was somewhat constrained. AM&M was sharing the development costs with Pratt & Whitney on a 30:70 basis.

The downside of the airbreathing engines was the lower thrust-to-weight ratio (T/W) and correspondingly heavier engines because you still needed the same maximum thrust.

The advantages of airbreathing engines for the landing phase had gotten Tom to thinking again about the tradeoffs for a high-altitude launch site. In theory, there was some advantage to launching a rocket at, say, 17,000 ft above sea level in the Peruvian Andes and at the equator. At 17,000 feet you're above half of the atmosphere, and thus drag losses would be substantially reduced, and the lower atmospheric backpressure would allow you to operate the first-stage engines at higher *Isp* for greater performance. However, operations at higher altitude, although improving the performance of a rocket-powered landing system, would have the opposite effect on an airbreathing landing system.

THE next time I talked to Tom, he went over the numbers that the performance group had generated. Given a vehicle optimized for high-altitude launch with a rocket-powered landing for the first stage, the payload gain would be less than 10%, making it doubtful that the logistics costs involved in flying from such a remote location would be justified. Sure, you could gain some advantage by flying from Denver at 5000 feet with rocket landing engines, but it would not be so significant that Denver was going to have any real cost advantage as a launch site for the DH-1. For an airbreathing landing system, high-altitude launch was a definite loser. Even at Denver, the landing engines would require thrust augmentation with water injection.

The first-stage test vehicle out at Mojave had the same mass properties and thrust level as called for in the current production vehicle configuration and was being operated under an experimental aircraft certificate. The only major differences between the test vehicle and the production vehicle were that the jet engines weighed about three times as much and had fuel consumption rates that were about twice the level that would be acceptable for the operational DH-1. For the test program they were using some surplus Boeing 737 engines (Pratt & Whitney JT8D-15 turbofans, derated to 12,500 lb), which had been mothballed when they couldn't meet the Stage 3 noise reduction standards and were therefore available and cheap. The test pilots started out gradually with short hops and touchdowns and worked their way up to "zoom flights" to 45,000 feet, where two of the engines could be kept lit while the other two were shut down, and then restarted on the way down. They had encountered some difficulties with the relights. After running one of the engines in a high-altitude test chamber, they figured out how to get reliable restarts at about 35,000 ft using TEB (triethyl-borane) injection. With that in hand, they were regularly "zooming" the stage to flameout at 50,000 feet, followed by relighting all four engines at 25,000–30,000 feet.

In addition to getting the engine restart system worked out, they were able to take care of the inevitable bugs that crop up in any new flight control system. The kinks in its vertical landing flight control system had made the jet-powered lunar lander trainer one of the more dangerous pieces of training equipment during the Apollo era. At the same time, the Mojave team was gaining some useful experience with ground handling and maintenance of the vehicle.

Chapter 19
Pilots, Payloads, and Passengers

WHEN I got back to Cape Canaveral, I took the opportunity to visit the high-fidelity mock-up of the orbital-stage payload bay/crew cabin. Jennifer Patterson, a self-described logistics engineer who was responsible for payload support systems, had called me to let me know that the mock-up was now available for a tour. Jennifer was there herself when I arrived, and we introduced ourselves. It turned out that after high school Jennifer had joined the U.S. Air Force, eventually becoming a loadmaster on C-141s during the Gulf War before she went back to school for a degree in engineering. Seeing that I was anxious to get a look inside, she led me over to the mock-up. "This is pretty neat," she said. "The rest of the vehicle basically exists for no other purpose than to haul this part, and its payload and passengers, into space and bring 'em back again. We'll use this mock-up for checking out the fit and function of all the payload and crew equipment, and for training the crews."

"I've been in the flight simulators already," I said. "Does this mock-up have those same functions?"

"No, we didn't see any reason to duplicate the flight controls here," Jennifer replied. "This is where we'll train the crews on how to use the payload handling equipment, and other equipment such as life support. We'll also train the ground crews on how to install everything and configure the interior for different missions."

I peered inside through a large, 6 × 6 foot open hatch that was, Jennifer explained, the main payload bay door. The volume inside reminded me of one of those double-concave lenses because the oxygen tank above and the hydrogen tank below formed roughly hemispherical cutouts in the cone frustum that constituted the payload bay section between the two tanks. I was somewhat surprised to see that the interior walls were smooth, with no signs of any stringers or bulkheads. Then I remembered what Tom had told me about the design of the tanks and payload bay. Like the tanks, the payload bay walls were constructed of 1/10-inch-thick aluminum skins that were capable of supporting even the fully loaded oxygen tank up above. During launch, a slight overpressure above the nominal atmospheric pressure in the payload bay would keep the walls in tension and allow them to support the loads experienced by the vehicle during first-stage flight.

In the center of the bay was a post, which Jennifer explained, consisted of the oxygen "downcomer," an insulated pipe that carried the oxidizer down to the engines, and several other vent and pressurization lines. It proceeded through a central hole in the hydrogen tank to the engine compartment. The hydrogen tank dome itself was covered with a lightweight, quilted insulating blanket stretched over its surface, except where the dome's stiffening rings formed a circular, double-rail track extending all of the way around the bay. There was a similar insulation blanket and a pair of stiffening rings on the oxygen tank; those rings formed another track overhead. The tank domes were actually semi-ellipsoids with axis ratios of 1.414:2:2, and so required the stiffening provided by the aluminum rings, which were welded in place. To the aluminum rings were bonded lightweight, composite insulating rings, which served to minimize heat flow from the payload bay into the hydrogen tank.

My attention was drawn to a gleaming open lattice structure that looked to be about 2½ inches in depth. It extended from the exterior wall to the hydrogen dome, about a third of the way up from where the payload bay wall blended into the tank wall, to form a deck of sorts. "That deck is quite a piece of hardware," I remarked. "It's beautiful!"

"Yes, it is—plated with gold, actually. Not something you'd see in a typical spacecraft, I suppose," said Jennifer, shaking her head with a smile. "Frankly, I think it's a little flashy, but Mr. Forsyth wants this vehicle to have some eye appeal—as if a real low-cost reusable launch vehicle isn't appealing enough! But he's got this thing about making this ship as attractive as humanly possible to potential customers. In addition to form-and-fit checks and crew training, this mock-up also serves as a marketing tool. Even without the gold, though, that deck is pretty neat. The upper portion is beryllium, brazed to a magnesium lower structure, which results in a very stiff and strong, but exceedingly lightweight deck. It helps to have something to walk around on

"Mr. Forsyth wants this vehicle to have some eye appeal—as if a real low-cost reusable launch vehicle isn't appealing enough!"

when you're on the ground, and you can also attach things to it on-orbit." Beneath the lattice, the hydrogen tank dome and the external wall of the payload bay formed a narrow, roughly triangular toroidal-shaped space. Jennifer pointed out that this volume was used to mount compressed air bottles, hydrogen-peroxide tanks, water containers, and other consumable supplies.

On the far side of the cabin, opposite the payload door, was the pilot's ejection seat, mounted on rails pointing toward a long window, which, I realized, would be overhead when the vehicle was in the landing mode. In the launch configuration, the pilot would be on his back, but during landing he'd be in an upright position. Because of the conical shape of the orbital stage, this window provided visibility forward as well as straight out, or up. Jennifer told me that the initial production vehicles would have only the window over the pilot's station, another small one in the payload bay door, and one in the emergency exit hatch. However, for additional visibility during critical maneuvers a string of lightweight digital video cameras would be mounted at 20-degree intervals around the exterior upper periphery of the cabin. The pilot could rapidly switch between these cameras or view a panorama on his main panel display as needed for use during rendezvous, docking, or landing.

Because the payload door could not be opened while the cabin was pressurized, there was a smaller emergency exit hatch, located at 90 degrees around the circumference from both the large door and the pilot's seat. This hatch could be opened under pressure by means of an explosive charge. An explosive-shaped charge would also be used to cut an opening in the wall adjacent to the pilot's seat, providing an exit for the ejection seat and pilot.

The pilot's seat, ejection rails, and control consoles were all mounted together between the two circular tracks on the tanks, so that they could be easily installed or removed. Jennifer pointed out how this technique would make it easy to add additional pilot or passenger seats. She told me that the current design goal was to be able to accommodate up to a total of five ejection seats. Next to one possible passenger ejection seat position was a painted ring, indicating where an adhesive-mounted explosive-shaped charge would open an additional exit hole for that seat. Eventually, a total of 20 crew and passenger couches could be installed, once the vehicle's safety and reliability record justified removing the ejection seats. The cabin was designed to have as many as five additional small viewing windows, symmetrically spaced about the cabin circumference, for passenger use.

I noticed some other rather luxurious appointments here and there—more of John Forsyth's "eye appeal." Jennifer explained that the use of precious metal plating and very thin coatings of fire-retardant-treated wood and leather provided the interior of the cabin with a look that was a mixture of "Jules Verne" spacecraft, luxury yacht, and high-end business jet, with very minimal weight penalties. Forsyth had an eye for aesthetics, and he knew what he liked. He had searched long and hard until he found an industrial designer who could produce a look that communicated high-quality attention to detail, while appearing futuristic and retro, all at the same time. "A lot of different kinds of people are going to have to get really motivated to bring together a lot of cash, if our sales targets are going to be met, and it can't hurt if the crew cabin is very attractive inside," he had told Jennifer. In fact, Forsyth had encouraged the industrial designer to work with a luxury car company to bring out a new high-end sedan, which incorporated some of the design features of the orbital stage. The ad campaign for the car would feature the orbital stage, communicating an image of progress associated with the space vehicle. And there was always the possibility of using the luxury car as a premium for DH-1 salesmen and purchasers alike.

Of course, as Jennifer reminded me, the vehicle was designed to accommodate cargo as well as passengers, and would likely be carrying more of that for some time. She proceeded to walk me through the mock-up, pointing out the various payload interfaces and support systems. The central components were the wedge-shaped cargo containers that were designed to fit through the payload bay door and slide onto the rails or circular tracks mounted on the two propellant tanks. They could then be easily maneuvered into positions so that the mass of all of the containers for a particular mission would be balanced about the central axis of the vehicle. Inboard of the large rails was a second set of smaller-diameter rails surrounding the downcomer. This second set of rails, Jennifer explained, would be used to mount much smaller payload containers, which might hold denser, bulk payloads such as water or hydrogen peroxide. The mocked-up containers were not unlike standard aircraft containers. It seemed to me that they would add unnecessary weight, and I asked Jennifer about that.

"The basic intent is to reduce the cost of payload integration," she began, "by allowing the customers to pack their payloads into standardized containers. Each loaded container will be inspected and its mass properties determined. Then it will be subjected to three times the maximum launch and abort loads. If the payload doesn't shift or damage the container during the loads tests, the mass properties—mass, moments of inertia about all three axes, and the coordinates of the center of mass of the loaded container—are plunked into some software we've developed that will integrate the mass properties of each loaded container with those of any other containers and spit out the optimal positioning for each of them. That means optimal from the perspective of overall vehicle balance, stability, and control. The containers will then be positioned along the tracks at their designated positions." She showed me markings on the payload rails that corresponded to witness marks on the container clamps, thus making it easy to properly position each container.

REMEMBERING what I had read and heard about the time-consuming and costly cargo integration process that was required for space shuttle payloads, I said, "That sounds so simple. Is that all there is to it?"

"Pretty much," replied Jennifer. "We've got some restrictions and additional requirements for potentially hazardous substances or devices and the like, but if the payload and the container hold together, and we know where to load them, that's all we need to do. And it really has to be that way, for low-cost space transportation. So, in answer to your earlier question, yes, the containers do add some weight, but they also greatly simplify the task of manifesting cargo, just like they do for air cargo." But as she pointed out, although air transportation costs are on the order of a $1 per pound of cargo for a transcontinental flight, transportation to orbit is likely to cost at least $100 per pound for the foreseeable future, and so there is additional incentive to use as light a cargo container as possible. As a result, the AM&M design utilized advanced carbon composites sandwiched between beryllium skins. "Of course, there will be some special-purpose payloads, such as larger satellites, that will attach directly to the rails. Depending on just how big they are, they may have to be separated into two pieces so that they can be properly balanced about the central axis of the vehicle."

"So how do you deploy the payloads on orbit?" I asked.

Jennifer led me over to the large door. "Well, the pilot will first depressurize the cabin, then open the payload door, which as you can see here, breaks in the middle, and folds up and down against the exterior wall. Then he'll unclamp a payload container, drag it along the rails to the open door, open it, remove the contents, and if it's a simple, self-contained satellite or spacecraft, simply push it out the door, with some velocity to make sure it moves away from the vehicle. For other, more demanding satellites, a lightweight truss can be erected and attached to the doorframe, extending up to 15 feet outward. Any payload that requires more precise deployment can be launched from the truss, with the vehicle oriented according to where the customer wants his payload to be pointed. If necessary, a spin table on a bearing race can be mounted on the truss, and the pilot can use a small hand crank to spin up the spacecraft. Then the entire spin table can be slid forward to give the spacecraft a small forward push. If a payload needs to be assembled, the truss provides a work area. That will allow the erection of larger or more fragile spacecraft that can't be launched in one piece. For instance, if a large antenna is needed, the satellite can be slid out onto the far end of the truss, where you have plenty of room to erect the antenna and attach any other parts of the spacecraft. At some point in the future when cargo is being transported to a space station, or to another DH-1 or other spacecraft, the truss could be used to dock the vehicle to the other craft."

"For other, more demanding satellites, a lightweight truss can be erected and attached to the doorframe, extending up to 15 feet outward."

During my tour of the cabin, I had noticed a couple of spacecraft mock-ups positioned on the rails. "Whose satellites are those?" I asked.

"Those are some new ones that we've been helping the customers to design," said Jennifer. "You should go talk to our satellite design group."

"I didn't know AM&M had one," I replied. "I figured everybody was busy enough designing the DH-1 without worrying about building satellites."

"Actually, they don't build them, they just work with several small satellite companies; and it's only three people. But one of the investors, who made a good bit of his money in the satellite-building business, came up with the idea. The plan is to design some standard spacecraft, with or without prepackaged instruments, that can be launched during the later flights in the prototype test program, and during the flight test program for the first batch of production vehicles. That will generate some revenue and jumpstart the low-cost satellite business. They're even working on an upper stage system, or propulsion module, that will be able to send these spacecraft buses all over the solar system."

THE next morning I was in the office of the satellite design group talking with Dave Morton, head of the team. Dave was only a couple of years out of college, and had led a small satellite research group at Stanford University.

He told me about several of the instruments and payloads that were planned for deployment on the standard bus.

"What about the upper stages that Jennifer told me about?" I asked.

"Yeah, that's some stuff we've been working on in-house," he responded. "Very cool, if I do say so myself. Come on out into the shop and I'll show you what we've got."

The basic upper stage consisted of a one-stage solid-propellant rocket, which could be loaded with between 1000 and 2000 pounds of propellant; two of these stages could be stacked together for greater performance or larger payloads. A liquid-fueled instrument bus on top completed the assembly. The bus had two 100-pound thrust engines that burned propane and hydrogen peroxide. The solid stage had a respectable structural fraction of less than 6%, but basically it was a rather ordinary solid rocket. Dave got more excited as he told me about the bus propulsion system.

It was quite a clever piece of engineering, but nothing really high tech. The hydrogen-peroxide tank was pressurized initially to 200 psi with helium gas, which occupied about 20% of the volume of the tank. During operation, when the pressure had dropped by 30%, a long silver needle was automatically driven into the tank through a hydrogen-peroxide-compatible grommet, much like you'd stick a syringe into a medicine bottle. The silver needle catalyzed the decomposition of some of the peroxide into steam and oxygen, and when the pressure in the tank got back up to 200 psi the needle was automatically withdrawn, stopping the reaction. The propane tank was fitted with electric heaters to increase the vapor pressure to 200 psi. With this system, the liquid stage had a rather impressive *Isp* of about 280 seconds in vacuum and a structural fraction of less than 10%.

On launch, the combined upper stage was spin stabilized, and the hydrogen-peroxide engines in the bus could be fired at any time to control the direction of flight by inducing the spinning stack to cone and then reducing the coning angle around the desired pointing vector. The modular nature of the stage provided the advantage that, by offloading hydrogen peroxide, propane, and solid rocket propellant, the total propellant weight of the stage could be as low as 1500 pounds or as high as 5000 pounds. With the mass fractions and engine performance that it had, the stage could provide enough delta-V to propel the instrument bus and a payload of almost 1000 pounds to Mars' orbit or 100–200 pounds to Jupiter's.

"Mars and Jupiter!" I exclaimed. "Wow!" Then as I thought about it, I said, "But that's not a very big spacecraft, especially to Jupiter, is it?"

Dave said, "No, not in terms of the big NASA or ESA probes. But we can pack quite a bit of useful hardware into them."

I was fascinated as he went on to describe that hardware. The basic configuration for the instrument bus contained an optical system based on the Questar 3.5-inch telescope, which could switch between a wide field of view for navigation and f/18 optics for planetary views, with a high definition television (HDTV) resolution sensor placed at the focal plane. A radiation-hardened CPU, with 0.25 terabytes of low-power RAM and dual 10-terabyte hard drives, was mounted with the optics on a platform, which, during a nonboost phase, could be magnetically levitated using pyrolytic graphite and neodymium-boride supermagnets, and then despun.

A laser communication system, using the same 3.5-inch telescope, would provide high data rates, while a microwave system using a small dish antenna would be available for low-data-rate transmission. For electrical power, a solar concentrator would feed an array of high-efficiency solar cells to provide up to 50 watts out to the orbit of Jupiter and over 1 kilowatt in near-Earth orbit. By deploying a larger, inflatable solar concentrator structure, a few watts of power could be supplied to the spacecraft as far out as Pluto and beyond. The idea was to launch a swarm of these spacecraft. Mass produced by one of the small spacecraft companies on an assembly line at a rate of 20 to 30 per year, the estimated cost was about $1 million per copy. Costs for the upper stage consisted primarily of the cost of one or two solid rocket motors plus the hydrogen-peroxide/propane engines and the guidance system; depending on the exact configuration, it was expected to be less than $0.5 million. Launching 10 to 30 spacecraft per year would eventually fill the solar system with probes that would check out moons, planets, and any number of interesting asteroids.

Launching 10 to 30 spacecraft per year would eventually fill the solar system with probes that would check out moons, planets, and any number of interesting asteroids.

The large memory capacity in the spacecraft allowed large amounts of data to be collected over long periods of time, and the high-data-rate laser system allowed large amounts of data to be transmitted over short periods of time. Dave's team had implemented a clever technique for compressing the data: their database was a "virtual globe" to which the data were mapped as it was collected, slowly building up a better and higher resolution map of the subject asteroid, moon, or planet. Also, the data would be layered, so that only the best would be saved. This database could then be queried, first by downloading a fractal model, or by randomly selecting areas of higher resolution. Then the globe could be queried to search for specific types of features. In this way, large amounts of data could be gathered autonomously and integrated for specific queries, which might involve a rather small subset of all of the data collected.

As Dave finished his discussion of the spacecraft features, I noticed an air table over in one corner of the shop, and something that looked like a small spacecraft—about the size of a volleyball—resting on it. We walked over to it, and Dave turned the air table on, handing me a remote control joystick. "Meet Wilson," he said, as if introducing a dog or a cat.

"Wilson?" I asked, raising an eyebrow.

"Well, as you can see, he is about the same size as a volleyball," said Dave, "as in Wilson volleyball. Get it?" He chuckled. "Take him for a spin."

I drove the small satellite around the air table with pulses from the tiny air jets that were mounted around the little vehicle. Then Dave told me to push the "Home" button, and Wilson whisked himself over to a docking port, where he was recharged with air as soon as he was connected. Dave explained that Wilson would ride to orbit with the upper stage, mounted in a niche in the exterior wall of the cabin. The small satellite would carry a multispectral camera and microwave transmitter/receiver and would be used to inspect the exterior of the vehicle while on orbit. With the microwave transmitter/receiver it could even measure the amount of water present in the heat shield to make sure it was good to go for reentry. Dave told me that he had worked on a similar spacecraft as a student at Stanford.

To minimize any damage that might result from a collision with the DH-1, the designers had kept Wilson's weight to just 5 pounds, and for good measure had surrounded the little craft with 1 inch of crushable foam. The benefits of having the ability to inspect the exterior of the vehicle far outweighed any risks that Wilson might damage the orbital stage.

As I was preparing to leave, I looked around at some of the engineering "trophies"—hardware prototypes—which were on a shelf next to the air table. One item caught my attention and I picked it up and asked Dave, "What's this?"

He explained that spacecraft typically require a complicated docking apparatus. In contrast, when docking a boat, a simple boat hook and ropes are used to maneuver the boat into its docked position. For spacecraft this approach would be difficult because you would need to depressurize the crew compartment before poles or ropes could be used to maneuver into a docking arrangement. One of Dave's engineers had come up with an idea based on a technique used on high-vacuum equipment to form a rotating vacuum seal around a shaft that penetrated a vacuum chamber. These devices used magnets in the bearings to hold a magnetic fluid to seal the shaft bearing against air pressure. Dave turned over the strange piece of hardware I had handed him and showed me a mechanism arranged like an iris in a camera shutter. This one was thicker, and the shutter blades were, he informed me, actually neodymium-iron-boron magnets. The opening of the iris was linked to a piston, so that as the aperture was enlarged by operating a lever, more magnetic fluid flowed into the perimeter formed by the iris; as the iris closed, the piston acted to remove excess magnetic fluid from the inner diameter of the iris. In this way a pole, or cable, or communications line could be passed through the wall of the crew compartment without depressurizing the cabin, thus allowing manipulation of objects or the spacecraft from inside.

I was intrigued and could readily see how a few of these ports, which seemed to weigh only about 10 pounds or so, could dramatically simplify the process of connecting a DH-1 orbital stage to a docked vehicle or an on-orbit supply of air, power, propellants, or data. Dave also told me that a slightly larger model was in test, which would allow an astronaut's arm to penetrate the spacecraft wall, much like the gloved sleeve in a glove box. He described how use of such devices could also be used to deploy satellites and close latching mechanisms. As usual, I came away impressed by the breadth of the innovative ideas that the engineers at AM&M were continually putting into practice.

CHAPTER 20
Mooncars, Monks, and Monasteries

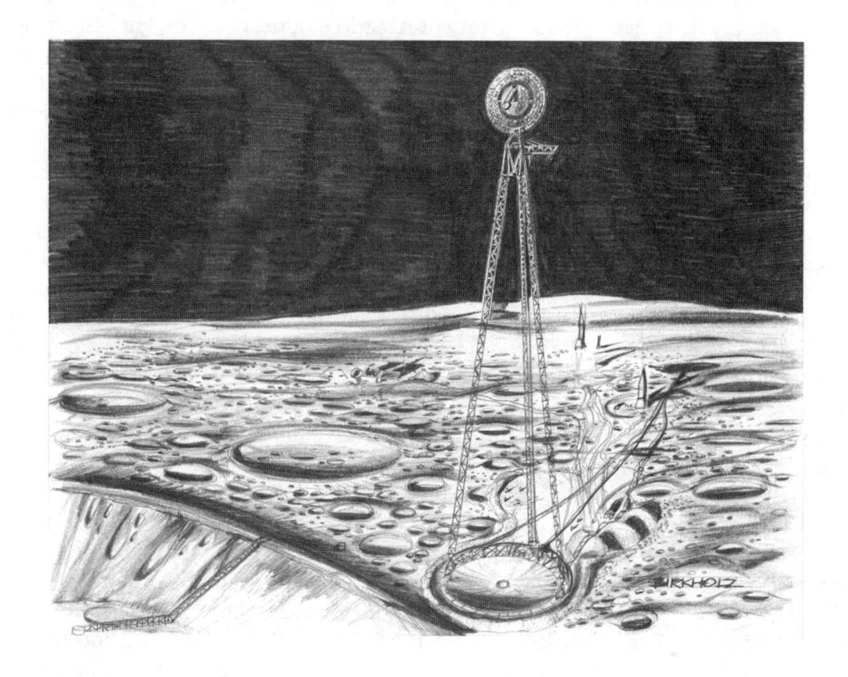

ALTHOUGH it was John Forsyth's policy to avoid direct investment in any outside company, from time to time a supplier needed to be encouraged or helped along with the development of a particular product or technology, usually with some technical assistance, but occasionally with some funding in the form of small contracts. He was also interested in encouraging potential customers, and, although he did not provide cash directly to such organizations or individuals, he made available a certain amount of money to fund internal studies and analysis that would support their concepts or projects. Usually, this involved investigation of modifications to the DH-1 or the logistics issues associated with particular missions. The idea was to aid prospective customers in their individual quests to raise their own money, so that they could come back and buy DH-1s. One of the most ambitious of these projects was being promoted by a New Zealander, Alexander Krempon, who had seven Everest summit successes to his credit. A seasoned explorer in both alpine and ocean environments, Krempon now wanted to go to the moon.

More particularly, he was planning an international translunar expedition, comprising 12 men and women each from a different country, who would circle the moon on the ground from pole to pole and back again. He had recently approached AM&M with a fairly well-developed plan. The moon is approximately 6000 miles in circumference, and Krempon intended to complete the expedition in about one year, traveling approximately 20 miles per day with a caravan of six vehicles. He had initially wanted to use fuel-cell-powered vehicles of new design, but after some lengthy discussions with a group of AM&M engineers led by Fred Clemens, he had been convinced to use instead an aluminum-framed, plug-in type hybrid-powered sport utility vehicle (SUV) from Japan—suitably modified for lunar operations, of course. Fred, who had worked in a variety of engineering jobs in an eclectic set of industries over a career of some 40 years, had the job of identifying and supporting projects that would help to create demand for the DH-1. I went to see Fred after hearing about the Krempon expedition during a session with Forsyth. I was quite intrigued by the idea of using an off-the-shelf SUV as a moonbuggy, and a little bit skeptical. Fred began by explaining why the SUV's power train was ideal for use on the moon.

I interrupted him almost at once. "I just assumed that the engine would be replaced with something that would be able to operate in a vacuum," I said. "But you're going to use the one that comes with the production vehicle?"

Fred smiled and replied, "Oh, it will take a certain amount of customization, for sure. But not really all that much, when you think about it. Let me start by reviewing how a hybrid vehicle works."

I had a general idea, but Fred made it clear. A hybrid vehicle has two powerplants: an electric motor running off of a battery pack and a relatively small gasoline-fueled internal combustion engine. The small engine can provide all of the horsepower required for constant-speed cruising—which is on the order of a mere 20 hp—but it doesn't need to provide all of the power required for acceleration or climbing hills. In those situations, the electric motor is used to provide a boost. This allows the internal combustion engine to operate at peak efficiency at all times because it doesn't have to do all of the work across the full range of power requirements. It is also used to drive a generator to recharge the batteries. The electric motor can also be employed to brake the car by running the motor backwards so that it acts as a generator, providing additional recharging for the batteries.

AS Fred explained, some relatively simple modifications would make the production SUV ready for lunar use. The first and easiest step was to fill the engine crankcase with Mobil 1 synthetic oil. This oil, intended for use in automotive engines, was actually a good high-vacuum oil and was often used in vacuum pumps. A high-vacuum oil is one that has such a low vapor pressure that even when exposed to high vacuum at normal temperatures virtually none of it will evaporate. The second step required a little more engineering—the standard wheels and tires had to be replaced with wire wheels similar to those used on the Apollo lunar rovers. Third, the springs and shocks were replaced with parts tuned to the higher mass but lower weight that the vehicles would have as used on the lunar surface.

The next task was to modify the gasoline engine so that it could run without an atmosphere to provide the oxygen for combustion. This did not entail much more than reprogramming the electronic control module to properly burn a mixture of oxygen and propane, diluted with carbon dioxide as a working fluid. The gas tank would be replaced with a liquid oxygen (LOX) tank and a propane tank. That was all it took to make the engine work in a vacuum. The coolant—your basic ethylene glycol/water mixture—would be run through one of two radiators on the left and right sides of the vehicle. Because the expedition would generally be heading either due north or south, one side of the vehicle would almost always be in shadow and thus better able to radiate the heat from the engine. A small pump would circulate the coolant past a small heater when the engine wasn't running, to keep it from freezing in the lunar darkness. Heat is lost by radiation as the fourth power of the temperature so that little power would be needed to keep the coolant from freezing at −40°F. The exhaust would run through a separate radiator, where the gases would condense to produce water that, after filtering, would be suitable for drinking, and CO_2, which would pass through an ion exchange column that absorbed most of it. The remaining CO_2 would serve as additional dilutant for the oxygen and propane burned in the engine.

The next task was to modify the gasoline engine so that it could run without an atmosphere to provide the oxygen for combustion.

The SUV's lithium-polymer batteries were scaled for a battery-powered range on Earth of 20 miles, which would give the vehicle a similar range in lunar conditions on battery power alone. The internal combustion engine and the electric drive motor were both somewhat underpowered for Earth use, but quite a bit overpowered for use on the moon. Aside from the greater than optimal weight of these powerplants for lunar conditions, the vehicle actually lost very little efficiency from operating the gas engine less frequently and the electric drive motor at less than full capacity. On the plus side, there was always power to spare for other uses. This meant that additional auxiliary equipment could be run off the electrical power bus without unduly taxing the capabilities of the generator. To minimize fuel use, the hood of the vehicle was covered with lightweight, high-efficiency solar cells, and a movable array was mounted over the rear seat/cargo area and could be moved from one side of the vehicle to the other to roughly track the sun. During the lunar day, these could provide almost all of the energy necessary to travel the planned 20-mile per (Earth) day stage.

The SUVs got about 40 miles to the gallon on Earth, which Fred told me amounted to about 7 miles per pound of fuel. Under lunar conditions, of course, oxygen had to be carried in addition to the fuel. The vehicles would also be carrying a mass substantially greater than their design payload on Earth and would be operating cross-country all of the way on the generally loose lunar soil. Taking all of that into account, Fred had calculated that they could be expected to achieve 1.5 miles per pound of LOX/propane on the lunar surface. When solar energy was available, very little fuel would be needed, so that each vehicle would consume somewhere between 0 and 13 pounds of propellants per day. The life-support provisions for the 12-person expedition would amount to about 5 pounds per person per day. Breathing air would be purified by removing CO_2 with ion exchange columns, and most of the drinking water would be provided by condensing the engine exhaust. Thus, the expedition would require between 60–130 pounds per day in consumables, including spare parts and other minor items.

Three of the vehicles were to be fitted with a lightweight habitation module installed in the rear seat/ cargo area. The other three were intended only for hauling supplies. The production vehicles had an Earth-side payload capability of 800 pounds. Fred had run some dynamic simulations that showed, for lunar conditions and a maximum loaded forward speed of 20 miles per hour, about 4000 pounds could be carried safely. Six vehicles with 4000 pounds each equaled 24,000 lb of payload. Fixed equipment, including storage vessels for the liquid oxygen, propane, and the habitation modules came to about 6000 pounds, leaving 18,000 pounds for consumables. That would be good for five to six months of trekking. The basic vehicles themselves weighed in at 2000 pounds each, for another 12,000 pounds, yielding a total of 36,000 pounds for the first half of the expedition. An additional 20,000 pounds of supplies would be landed at the north pole for the second half. Thus, the total weight that the expedition would need to have hauled to the moon was about 56,000 pounds.

"Given that the payload capability of the DH-1 is 5000 pounds to Earth orbit," I asked, "how do you plan to get 56,000 pounds to the lunar surface?"

"Remember, a fully loaded, fully fueled orbital stage has a delta-V capability of almost 25,000 feet per second just to get to orbit," he responded. "The total delta-V needed to get from low Earth orbit to the lunar surface is about 18,000 feet per second. That breaks down to about 10,200 feet from low Earth orbit to an orbit intersecting that of the moon, and another 8000 feet per second for direct descent to the lunar surface. Return from the lunar surface requires an additional 8000 feet per second, if you aerobrake in Earth's atmosphere. The total delta-V then is 26,200 feet per second. So a refueled orbital stage would be marginal for a flight to the lunar surface and return—it could deliver a couple thousand pounds to the lunar surface and return empty to low Earth orbit."

"However, after talking things over with the performance and vehicle engineering folks, we came up with a concept for a lunar transport that should be able to take 35,000 pounds to the lunar surface and return the vehicle to low Earth orbit." He paused.

I let out a low whistle. "That sounds pretty cool! Is it a new vehicle?" I asked.

"No, it's actually a pair of orbital stages, connected nose to nose," Fred stated, with a gleam in his eye, and looking rather pleased with himself. "Basically a two-stager. One of them, the lunar stage, will have a stretched payload bay—20 feet longer than the standard model—but we can still keep the empty weight under 17,000 pounds so that we can get it to orbit. Without any other payload, of course. We'll launch both DH-1 orbital stages to a refueling station at 150 nautical miles, hook them together and fuel 'em up, and launch the combination into an orbit just shy of 10,000 nautical miles, with a total delta-V of 6000 feet per second being provided to the lunar-bound orbital stage by the first one. That will allow the lunar stage to carry 35,000 pounds to the lunar surface, off-load it, and return to low Earth orbit. It'll use aerobraking to slow down enough to enter the refueling station orbit for rendezvous. That will become its home base—it will stay in space for future lunar runs."

ALWAYS aware of AM&M's emphasis on the cost issues, I asked, "So, how much will it cost per pound to the moon?"

Fred had already worked that out. "With the combo system, you get approximately 17% of the total payload launched into Earth orbit—that includes the propellants for refueling, and other miscellaneous hardware—to the lunar surface. So, if 1 pound of payload to Earth orbit comes in at $200, 1 pound to the moon would cost on the order of $1200. We will have to spend some additional money on the lunar vehicle, too, of course. In addition to the stretched payload bay, the lunar stage will require some special modifications for lunar landing, not to mention a heavier heat shield for aerobraking back at Earth. Actually, as it turns out, the heat shield won't be that much heavier. For aerobraking to low Earth orbit velocity, the energy dissipated will be about the same as for Earth reentry. The heating rate will be higher, but the length of time will be shorter, and for a rechargeable water-ablating heat shield like we've got, the total heat is the major factor driving the weight."

I did some mental calculation. At $1200 per pound, 56,000 pounds of supplies delivered to the lunar surface would cost somewhere in the neighborhood of $70 million. "Less than $100 million to take Krempon's expedition to the moon! And NASA is always talking about 'billions and billions.' "

"There will be some other costs," Fred interjected. "Stuff related to deep-space operations, added amortization of the modified lunar vehicles, related logistics needs. Overall, I'm estimating a total in the $200 to 300 million range. But, if we get costs to orbit well below $200 per pound, that total could come down quite a ways. Two-hundred-fifty million dollars is probably conservative, but it's a good estimate at this point.

"Here's one more modified vehicle we'll need." Fred showed me a drawing of a DH-1 orbital stage designed to boost one of the 2000-pound lunar SUVs to orbit where it would be transferred to the lunar stage. The payload bay would be stretched 4 feet, and the oxygen downcomer would be lengthened and given a U shape, extending first radially outward to the outer wall of the payload bay, then down along the wall, and then back into the center of the hydrogen tank. He pointed out a zero-spill, quick-disconnect valve located just below the oxygen tank. This would allow the entire oxygen tank to be swung away, providing access to the payload bay for off-loading the SUV, which would be carried in a lightweight cradle mounted to the payload support rings on the oxygen and hydrogen tanks.

FRED then asked, "By the way, are you aware of the lunar monastery project we've been working on?"

"Yes, John Forsyth mentioned it the last time we met. Interesting concept—certainly not one you usually see on the 'top ten things to do on the moon' lists. But I understand they're quite serious about it."

"Very serious. Investor Number 3 is willing to put up a billion dollars—that's serious in anybody's book. I've been spending quite a bit of time on it lately. As it turns out, Number 3 and Father Scipio will be here on Thursday for a meeting. Maybe you'd like to join us," Fred suggested.

"Yes, indeed. Would they mind, do you think?" I was a little unsure about meeting this "moon monk"—the term just popped into my head—but I was definitely curious.

"I don't think so, but check with Number 3. It's not hush hush or anything, but I know they don't exactly advertise it, either."

After leaving Fred's office, I decided to give Forsyth a call first. He gave me some background. It turned out that Number 3's primary motivation for backing AM&M was to be able to found a monastery at the lunar south pole, and that the planning for this had been underway for more than 10 years. Forsyth told me he would be seeing Number 3 that evening, and he'd pass on my request to learn more about the monastery. I got a call the next morning from Number 3 himself, with an invitation to the meeting.

Thursday morning I found myself in the conference room across the hall from Fred's office, sitting next to Number 3 and across the table from Father Scipio, a Benedictine monk of the Reform, as I was to learn. He was a small, muscular man, perhaps five-feet-five inches, in his late 40s or early 50s, and he wore the habit of a Benedictine. He chose his words carefully, but he had a presence and demeanor that was both reassuring and demanding, and when he talked to you, you could be sure that he was deeply interested in both you and what you were saying.

Number 3 had told me something of Father Scipio's past when I spoke with him the day before. He had been a mining engineer and shift boss in a hard-rock mine in Colorado when he heard the call to the religious life. Like some of the early church fathers who had felt called to live in the Egyptian desert, he had felt a special call to found a monastery on the moon. He had met Number 3 in Rome, while there tending to the early steps in the process necessary to found a new Catholic movement. Coincidentally—or perhaps not so coincidentally—Number 3 had come to Rome at that same time, looking for someone to build a monastery on the moon, for which he would foot the bill.

Like some of the early church fathers who had felt called to live in the Egyptian desert, he had felt a special call to found a monastery on the moon.

That was over 10 years ago, and ever since, Number 3 and Father Scipio had been quietly but energetically planning and working towards the lunar monastery. Father Scipio's new Benedictine order had grown to six avowed members and 12 novices in various stages of discerning their vocation. They were living in an abandoned mine in Colorado, in what had been an exploratory shaft driven horizontally into the side of an outcrop. Ultimately, this shaft had not shown a very promising ore body and had been abandoned. However, for Father Scipio's purposes it was ideal, situated as it was in a large body of fractured rock. Father Scipio was of the opinion that the rock body into which the shaft had been driven was actually the base of an eroded impact crater, and thus the rock properties, composed of fractured but not dislocated large blocks, would be representative of what might be expected a few hundred feet below the pulverized rock of the lunar surface.

Father Scipio and Number 3 had graciously agreed to meet me an hour prior to their scheduled meeting with Fred, and Father Scipio was more than happy to tell me about his order, their community in Colorado, and his plans for the moon. The monks and novices led simple lives of hard work and prayer—in typical Benedictine fashion, as Father Scipio explained it. Their work was focused on experimenting with and learning how best to open passageways off of the existing mine shaft. They had cut a small chapel, dining room, and individual cells out of the living rock. The techniques they used were simple and well known to those in the mining and tunneling industries. The major difference was that they were attempting to use electrically powered drilling jumbos, instead of those heavy-duty, vehicle-mounted jackhammers used in mining. The electric versions not typically used in mining were somewhat underpowered. But a manufacturer who had made great strides in increasing the effectiveness of their electric jackhammers wanted to expand the market for their new model in terrestrial mines and was working with the monks in real-life field trials.

Father Scipio proceeded to tell me about their technique for creating openings in the rock. First, a central hole is drilled where the opening is desired, starting at about 1½ inches in diameter. Then it is gradually expanded with larger and larger bits, until it's about 4 inches in diameter. An array of other boreholes, each 1½ inches in diameter, is placed around the central hole, at gradually increasing spacing. The small outer holes are then packed with high explosives, which are set off in a planned sequence. Rock is very strong in compression, but rather weak in tension, so that the initial result of setting off a high explosive in a borehole is relatively ineffective, as the rock around the borehole is compressed. But the explosions generate a compression wave that travels through the rock. When it reaches the large central borehole, the discontinuity in density between the rock and the empty hole results in the wave being reflected from the borehole surface. The reflected wave is a tension wave, and, as it builds up, it rapidly exceeds the strength of the rock, breaking away a slab in towards the borehole and creating a new free face. The compression wave reflecting off of the new free face in turn creates another tension wave, which breaks the rock successively back towards the source of the original compression wave. After a delay of perhaps 50 to 100 thousandths of a second, the explosives in the boreholes further out are triggered. The compression waves from these boreholes have a larger free surface to act on, and successive rings of explosions continue on out to the edge of the desired opening. Using this technique it is possible to remove, more or less accurately, a prism of rock of arbitrary size and about 3 to 6 feet in depth.

Frequent fault lines complicated cutting the openings, but sometimes the monks could take advantage of the fault lines to speed the work. Their work consisted mainly in drilling the holes for the explosives and removing the debris. Because of the relatively high strength of the rock, the drilling was tedious and hard. Removing the rubble was just hard. Their method differed from tunneling only in that they desired a more highly finished surface on the interior, and more precise removal of the unwanted rock, in order to shape the rooms and chambers according to an architectural design. In the newly excavated living spaces, the monks had proceeded to carve decorative designs into the walls. Number 3 had been deeply impressed by their artistry.

The monks produced all of their own food. The diet consisted mainly of vegetables, with some fruits such as strawberries. Protein was provided by a small flock of laying geese and a small fishpond full of tilapia. There were also a couple of milking goats, providing a limited amount of cheese. The vegetables were grown in an intensely farmed hydroponic garden, with somewhat less than 1000 square feet per member of the community. The garden, pond, and animals were enclosed within a greenhouse that was equipped with high-intensity lights for use during the winter. The monks used composting toilets and extracted nutrients from the resulting compost to supplement the chemical feed for the hydroponic garden. The whole system was not particularly economical, considering the use of the high-power electric lights, and they were having some trouble getting a variety of crops available at the same time. However, the monks were not fussy, and the food was filling and nutritious.

The overall system closely approximated the approach that Father Scipio was planning to take at the lunar south pole. He intended to locate the monastery on a "peak of eternal light," along the rim of Shackleton Crater. The chosen site, on a promontory that was nearly always in sunlight, was one of several such identified by Weerd, Kruijff and Ockels from the Clementine lunar probe data just before the turn of the century. The monks would strip the ridge top of all easily removable rock and construct a greenhouse enclosing a little more than one-tenth of an acre, or about 5000 square feet in the center of a depression and surrounded by a circular track. The greenhouse was to be constructed from a number of carbon-fiber-reinforced plastic modules each weighting about 5000 pounds, which would be linked together. To keep the structural weight of the modules down, the total pressure in the greenhouse would be maintained at about 3 psi, with about 1 psi from water vapor and the other 2 psi from a mixture of carbon dioxide and air. By maintaining an even temperature inside, they would be able to keep the water vapor pressure very nearly constant. The high carbon-dioxide content meant that photosynthesis in the food plants would not be limited by the availability of CO_2. Webb suit-type spacesuits would allow the monks to work inside the greenhouse without prebreathing.

He intended to locate the monastery on a "peak of eternal light," along the rim of Shackleton Crater.

The air and carbon dioxide would be supplied from the monastery underground. A specially designed turbine would use the flow of air coming up from the monks' living quarters—which was to be at a Denver level of about 12 psi—to recompress the air from the greenhouse for recirculation through the underground living quarters. The compressor would be assisted by an electric motor to make up for losses in the turbine and the compressor itself. The recirculation would allow the plants to use the CO_2 produced by the monks and to replenish the oxygen in their breathing air. With 24-hour lighting, Father Scipio expected to get rapid plant growth and adequate purification and removal of excess CO_2. Even so, the air in the living quarters would be at a higher-than-normal carbon-dioxide content. Men can easily adapt to a carbon-dioxide concentration of 1 or 2%. After all, the air exhausted from a typical high-rise office building is often 0.5% CO_2. People can even adapt to levels as high as 5%, given sufficient time.

On the track surrounding the greenhouse, there would be a mobile tower, projecting up 300 feet into the nearly perpetual sunlight. At the top would be an array of mirrors, designed to reflect the useful visible light and a certain amount of the infrared, but not the ultraviolet, down into the greenhouse, which would be shielded from direct sunlight. The light would be at approximately one-third the intensity of the midafternoon sun in a temperate climate, but it would be available 24 hours a day from the mirrors, which would have a total surface area of about 1500 square feet. A solar-powered electric motor would drive the tower around the circular track every 27 days, which is the sidereal period of the moon. Because the mirrors would not reflect the ultraviolet part of the spectrum, which is deleterious to polymeric materials, carbon-fiber-reinforced plastic modules could be used for the greenhouse modules.

Initial plans were to construct the majority of the habitat in underground tunnels. That way, the overlying rock would serve as the pressure vessel to contain a breathable atmosphere. On Earth, for each foot you dig down, the lithostatic pressure load caused by the overlying rock is about 1 psi; it would be 15 psi at approximately 15 feet down. On the moon, with its lower gravity, a pressure of 12 psi would require a minimum depth of 75 feet. However, Father Scipio intended to go down to 200 feet, so that the rock would remain always in compression, thus minimizing the air lost to leakage.

Eventually, Father Scipio planned to move the entire agricultural area underground, where higher pressures could be sustained and the farming could be done without special suits and breathing gear. To provide natural lighting, he was going to drive a 12-foot diameter shaft down to the tunnels. It would be lined with highly reflective multicoated aluminum, so that even with an average of 10 reflections as the light bounced down the shaft, less than 20% of the light entering the shaft would be lost. At the bottom, there would be a 13-foot diameter quartz window in the shape of a parabolic meniscus, loaded in compression by the air within the underlying agricultural chamber. Averaging about 3 inches thick, the mirror would weigh almost 5000 pounds. It would have to be manufactured on Earth and shipped to the moon. In fact, in order to have a spare on hand, Father Scipio wanted two of them. The mount for the window could be cast from nickel-iron on the moon. The gaskets, however, would have to come from Earth.

Overhead, the mobile tower-mounted mirror system, enlarged to some 3000 square feet, would now be used to illuminate the shaft instead of the surface greenhouse. Sunlight concentrated to 30 times normal solar incident, minus the short ultraviolet and far infrared, would pass through the quartz lens and diffuse into a 10,000-square foot chamber to provide optimal growing conditions for the crops that would provide both more food and more air recycling capacity. With nearly 300 kilowatts of solar energy passing through the window, the ultrapure multicoated quartz was expected to absorb somewhat less than 1%, or about 3 kilowatts. This meant that 3 Btu per second would have to be removed from the window which would be accomplished by blowing air against the lower surface of the window. Filtering the cooling air would aid in keeping the window clean.

Initially, some 30 kilowatts of electrical power for basic needs would be provided by a 1000-square foot array of high efficiency solar cells mounted on the mobile tower with the mirrors. Additional power for extracting water, metals, and other useful materials from the lunar rock and soil would come from another solar electric system located in a notch in the side of the perpetually illuminated crater lip. The notch would be lined with reflectors concentrated on a solar voltaic array that could supply 100 kilowatts. The average radiation incident on the array would be three times solar normal and would be available nearly constantly over a 3-day period, with lesser amounts for the 3 or 4 days before and after, as the sun "rotated" around the sky over the lunar south pole.

The most readily available construction material on the moon is nickel-iron from the meteorites that have been impacting the surface of the moon for millions of years. This metal can be gathered simply by passing a magnet over the surface.

The most readily available construction material on the moon is nickel-iron from the meteorites that have been impacting the surface of the moon for millions of years.

Roughly 0.5% of the lunar regolith, or soil, is composed of meteoric iron. High-strength neodymium-iron-boron magnets would be incorporated in a surface mining machine not unlike a snow blower, with a harrow that would scrape material from the top several inches of the lunar soil. The machine would sieve out the small particles, pass the fine-grain material over the magnets to collect the nickel-iron, and dump the detritus. In this way, about 20,000 pounds of raw nickel-iron could be extracted per acre from the top 2–3 inches of lunar soil. An electric arc furnace, powered by the solar-voltaic systems, would be used to cast major structural elements for the monks' habitats, including the pressure doors and structural supports for the tunnel walls. They were experimenting with a waterless sand casting technique, using simulated moon dust in a small electric arc furnace at the monastery in Colorado. Although there were some bugs yet to be worked out, the system seemed to work for casting basic structural parts.

I was fascinated by Father Scipio's rather matter-of-fact description of his plans for the lunar monastery. It was quite clear that he knew exactly what he was talking about. I had never met an engineer-monk before, and I was impressed, both by his technical competence and his quiet earnestness. I had to ask him one question as Fred walked in to join us.

"Father, this is quite a plan you have," I began, and glanced from him to Number 3. "And I am impressed. But do you think the cost is justified? I mean, a lunar monastery will cost many times what one on Earth would cost."

"Yes, you are right about that," he said after a short pause. "But Number 3 has generously earmarked his own money for this project. We have both felt called to do this, and it seems nothing short of providential that he and I, and AM&M, have found each other at this time. As you consider the history of earlier eras of exploration and expansion, you'll see that the Church has always been there right from the beginning. We have no doubt that Mankind is poised to move from this one small planet out into the vaster universe, into the rest of God's creation. The Church must go, too. And yes, that means spending money. We have a mission wherever our brothers and sisters are found. Besides, just as the shrines and cathedrals of this world have stood as witnesses of faith and inspiration to all, a monastery and church on the moon, making its way across Earth's night sky month after month for everyone to see, will truly make it serve as a light to the world."

Fred settled into his chair and brought the meeting around to the topic that Father Scipio and Number 3 had come to discuss. They were here for an update on the estimated costs for transportation to the moon and to discuss methods for hauling the larger pieces of equipment. Like the hybrid SUVs that would be used for the translunar expedition, the monastery was going to use gasoline-powered tractors and front-end loaders converted to burn propane and oxygen. They would also need to get the two 5000-pound windows to the moon. The last time Father Scipio had been to AM&M, they were still talking about using single, refueled orbital stages to transport a few thousand pounds to the lunar surface. Fred wanted to fill him in on the concept for the two-stage version, which had been developed with the Krempon transpolar expedition in mind.

EVEN though Number 3 had committed a sizable sum to the venture, Father Scipio was keenly interested in exactly how much transportation to the moon was going to cost, as well as the other details of the logistics system that would need to be in place. Unlike the expedition, the monastery would require long-term support. The current estimate put the total materiel requirements right at 100,000 pounds for the initial 15- to 20-man monastery. Father Scipio and the monks had been continually adding detail to their list of necessary supplies, and their weight estimate was getting more accurate by the day. Based on their work in Colorado, Father Scipio had recently estimated that 2.5 pounds of explosives would be required for each cubic yard of rock and soil to be excavated. At 100 cubic yards per member of the community—for a total of 2000 cubic yards—including a small machine shop and repair area, over 6000 pounds of explosives alone would be required. But there were still a number of areas where more work was needed.

As Fred, Number 3, and Father Scipio laid it out, the necessary logistics infrastructure would include one lunar transport, modified from a basic orbital stage, at $150 million and an on-orbit refueling facility with a capacity of some 164,000 pounds of propellant at $200 million. That made it $350 million for transportation and support. Realistically, from a safety perspective, it would be desirable to have two lunar transport craft available, so that would make it $500 million. Number 3 was hoping to save as much as 30–40% on the lunar transports by negotiating a sweet deal during the next round of financing, thus bringing the cost down to about $400 million for the transportation systems. Father Scipio was figuring that the propellants and the services of another orbital stage to provide the boost to lunar transfer orbit could be purchased from some future low bidder, once the DH-1 was established in service. He therefore anticipated that the marginal cost for a pound delivered to the lunar surface might fall to as low as $600. So out of the $1 billion that Number 3 had committed, $400 million for the spacecraft, $250 million for the development of the monastery itself, and another $250 million for transportation, at $1200 a pound, of up to 200,000 pounds of personnel and supplies would leave some $100 million in reserve to cover contingencies or higher transportation costs during the first few years. Once the monastery was firmly established, it should be nearly self-sufficient, aside from sophisticated manufactured goods, nitrogen, and some other minor nutrients that would have to come from Earth.

I came away from the meeting deeply impressed by both Father Scipio and Number 3. These men certainly knew what they were doing, and were intense in—how can I say it?—in a very reassuring way. I was no longer skeptical about a monastery on the moon, and I no longer thought of Father Scipio as the moon monk. I was sure that if anyone could build the first permanent base on the moon, then it would be him.

Shortly afterward, I spoke with Forsyth again, and he told me that he found the monastery project very encouraging; in fact, he was quite enthusiastic. Each 1000 pounds delivered to the moon would require more than one DH-1 flight to low Earth orbit. This meant that if 200,000 pounds of payload went to the moon over 200 flights into low Earth orbit would be needed. With additional flights for setting up the refueling station and with Krempon's circumlunar expedition, it would take 50 to 100 flights per year simply to provide and support transportation to the moon. That was the beginning of a strong market, he had said, a market that could only be served effectively by the DH-1. An integral feature of both of these projects was the intention to buy launch services from other providers. That would provide an attractive market for prospective DH-1 customers, giving them a good reason for buying.

Beyond that, the availability of shelter at the lunar south pole would encourage scientists, explorers, and other visitors who wanted to go to the moon. The 10,000-square foot greenhouse should be capable of supporting more than 40 people, and, as the initial community of monks was only going to be 15 to 20, they would be able to accommodate up to 20 or so visitors. Eventually, Forsyth expected the existence of the monastery would lead to the establishment of scientific stations in the general vicinity, and thus a small lunar settlement would begin to develop. The south polar region was probably the most attractive real estate on the moon. Hydrogen was known to be relatively abundant there, down in the permanent shadows of the craters, in the form of solar hydrogen and perhaps even water ice. Father Scipio certainly intended to construct a solar still to extract this precious resource, which would provide water for drinking and for crops, as well as propellant for the spaceships and ground vehicles.

And, as Forsyth pointed out, a monastery on the moon with room for more than 20 visitors meant that there would be someplace for tourists to go and to stay on the moon—if not a lot of entertainment in the conventional sense. Of course, I had read those prognostications about "flying" inside a dome in the moon's low gravity, and there would be all of the natural wonders of a truly unexplored world. Yes, people would want to go. He told me that a single, conventional orbital stage, refueled in low Earth orbit, could probably transport 10 people to the surface of the moon and back, for approximately 15 to 20 times the cost of an orbital flight. Thus, if the cost of an orbital flight came out at $1 million, a round-trip to the moon might be around $20 million, or $2 million per ticket. Dennis Tito had paid $20 million just to spend a week in the International Space Station. And it didn't seem that Father Scipio intended to charge room and board at the monastery. You could probably stay basically as long as you wanted, especially if there were regular flights to and from the moon—although after talking with Father Scipio I could imagine that he would put you to work if you stayed long enough. But I had also gained the strong impression that staying in his monastery would be a thoroughly enjoyable and satisfying experience. And if there were regular flights, someone would come along and build better accommodations with more amenities and more entertainment, all fueling a growing market for space transportation and driving the demand for more DH-1s.

Except for people, there wouldn't be much traffic in the Earth-bound direction anytime soon. However, Forsyth had suggested to Father Scipio that a very small vintage of wine could be made on the moon and shipped back to Earth as a rare novelty item. Some terrestrial wines did sell for over $10,000 dollars a bottle, and he was always looking for something that would stimulate interplanetary commerce. But John Forsyth wasn't intending to wait around until the demand was there. He was going to develop the demand by careful nurturing the expectation that things like lunar expeditions would soon become paying propositions.

CHAPTER 21
Aliens, Cheetahs, and Archea

FRED invited me back the next day, when he promised to explain how the DH-1 could be used for Mars expeditions. After hearing about the plans for the moon, I could not wait to hear about Mars. I arrived at the AM&M building a little early, around lunchtime, so before our meeting I headed over to the cafeteria for a quick bite. As I looked for a place to sit, I came upon one of those episodic conversations that reoccur from time to time in any organization that is involved in any significant way in the business of space travel. The topic was, as Fermi had put it, "Where are they?"

"They" are the aliens, the extraterrestrials. The big question is, are we alone in the universe, and, if we're not, then why haven't they visited us? Now AM&M's workforce was composed strictly of the most competent talent that money and an exciting project could pull together. But they could be divided roughly into two camps—those for whom the "conquest" of space was a cause or a calling and those for whom it was a job—to be sure, a very exciting job, which they were proud to hold, but a job nonetheless, not all that different from building nuclear submarines, commercial airplanes, supercomputers, or even automobiles. Interesting work if you could get it, but certainly not a cause. To be sure, among the Group of Seven, those to whom space was a cause were decidedly in the majority. However, in terms of the working engineers and support staff, they were definitely in the minority.

Thinking of your business as a cause was nothing new. Certainly Apple Computer, in its early days under Steve Jobs, had been driven by a "messianic" mission to bring power to the people through personal computing. And working at Apple during those early years had tended to leave a lifetime impression on those who had participated. Viewing a company or a job as a cause had its advantages and disadvantages. Without a doubt, Apple had been way ahead of its rivals in both functionality and in seeing where the future of personal computing was headed. On the other hand, a myopic concentration on a specific vision of the future could blind an entire organization to reality in general and to the reality of the marketplace in particular. True, if the Macintosh had been made backward compatible with the Apple II, it might not have been as exciting a machine. But backward compatibility and an open architecture eventually allowed Intel and Microsoft to dominate. Likewise, in the software business Microsoft's more pragmatic approach allowed it to borrow ideas from competitors, rather than being fanatically dedicated to an exclusively graphical interface and a one-button mouse.

Although building the DH-1 was a cause for John Forsyth, he had been in the business world long enough that success was more important to him than filling in the exact outlines of a particular dream. Elizabeth Weil's book *They Laughed at Christopher Columbus*, which chronicled one unsuccessful launch vehicle company from the turn of the millennia, had also helped to open his eyes to the fact that building rockets—especially if "vision" dominated the enterprise—could seem to be irrational from an outsider's point of view. Of course, as the title of her book suggested, Weil was aware that the single-minded dedication of great men who accomplished great things often made them seem irrational to the rest of the world. Sometimes perhaps they really were irrational in their pursuit of an idea or a cause. It could certainly be said that all strongly held beliefs, be they religious or otherwise, to the extent that they were true and did not conflict with reason, went beyond rational thought to faith.

It was John Forsyth's hope that, aided by the clearer headed rationality of Tom Rabbet, that AM&M was sticking to the solid middle ground between the dreamers who would change the world and the realists who could lay a solid foundation for success. The company was decidedly dominated by the realists in numbers, but this didn't mean that the dreamers, with the power of their ideas, did not sometimes get the upper hand. From the dawn of the Space Age, progress had been by both the dreamers and the practical men. And in the last 20 to 30 years the companies founded by the realists had not made significantly more progress, in terms of opening space as a real frontier, than those founded by the dreamers—perhaps less.

THIS interesting mix of practical men and accomplished dreamers resulted not infrequently in interesting exchanges of views—and sometime heated debates—which broke out from time to time among members of the team. And nothing could provoke quite as lively a discussion as the topic of extraterrestrial intelligence. For those who had grown up on science fiction, the idea that there were other intelligent beings in the universe and that one day we would meet up with them seemed highly likely, if not a certainty. For most of the realists, the existence of extraterrestrial intelligence was not something that they really cared to spend much time thinking about. Undoubtedly, if aliens showed up on our doorstep, there would need to be some consideration of what to do about it. But in the meantime, it seemed like the farthest fantasy, somewhat akin to speculating about time travel and what would happen if you went back in time and killed your grandfather. Thinking too hard about it left one unsettled without any real benefit. Thus, it seemed, endlessly speculating about the possibilities of extraterrestrial intelligence only meant that you had too much time on your hands.

And nothing could provoke quite as lively a discussion as the topic of extraterrestrial intelligence.

So when I came upon a table full of engineers in the midst of a rather energetic discussion, it was one more round in the ongoing conversation. As I sat down to join them, Thor Thingveld, the chief metallurgist with the structures group, was explaining his theory about how the whole concept of SETI (the Search for Extraterrestrial Intelligence) was mixed up. Apparently, it was in response to some comments made by Amanda Larson, an electronics engineer doing some advanced planning work for investor Number 5, who hoped to finance a large SETI radio telescope array at the Earth–moon L5 point. Thor put forward his opinion that SETI was really a manifestation of pantheism, the belief that the universe was God.

"A heretical Jesuit priest named Pierre Teilhard de Chardin developed a theory in the first half of the last century that provided the underlying philosophical basis for SETI," Thor stated. "He was one of the chief practitioners of that twentieth century brand of theology called modernism that wanted to explore Christian beliefs in terms of certain philosophical ideas that arose in the nineteenth century, particularly the concept of evolution. He, and many others, viewed evolution not merely as a useful system for organizing and classifying organisms, but as the very underpinning of a new philosophy.

"Like 'liberation theology' in the mid-20th century, which tried to reconcile communism and Christianity, modernism was ultimately a failure and condemned by the Church. Even so, elements of these philosophies live on in popular belief. De Chardin was particularly influential in developing the concept that we were evolving towards God, with chemical evolution leading to biological evolution leading to intelligence in the individual, and then individuals being linked together into larger and larger intelligent societies. Finally, the whole globe becomes one unified intelligence, which he called a noosphere. Then the noosphere of the Earth will link up with all the other noospheres in the universe, composed of other intelligent species around other stars, so that the universe will become intelligent, and therefore become God. Of course, the basic Christian doctrine that God appeared as a man, in a very definite revelation some 2000 years ago, isn't exactly compatible with the concept of a universe evolving towards God. But de Chardin's ideas still provide a convenient basis for the modern pantheist. More particularly, it implies that our function as an evolving intelligence is to link up with other intelligences in the universe."

Amanda vehemently denied being a pantheist. "I've just always thought it would be really neat to meet other races and learn how they live and think. And, with the array that I'm working on, we should be able to eavesdrop on civilizations out to 1000 light years, so that even if they aren't beaming any messages at us, the normal leakage from their internal communications should allow us to listen in and learn all about them. For sure, the list of stars with known planetary systems keeps growing all the time! If they're out there, and at all commonplace, this project will find them."

Joe Forsberg from the reentry systems group interjected, "But what if they're already here?"

At this, everyone jumped on him, demanding evidence that aliens had ever visited the Earth. Joe responded with three examples, two of which at least hinted that Earth had been visited in the relatively recent past. The first, he explained, was the cheetah.

"It turns out that cheetahs exhibit no genetic diversity. They are so similar that a skin graft between any two cheetahs will take. They are practically all identical twins." Joe paused, looking around to see if anyone knew what that meant. "That," he pronounced somewhat triumphantly, "is biologically impossible. Even if the cheetah population had been reduced to one male and one female, very little genetic diversity would have been lost. The only way to account for the lack of diversity would be to have very few cheetahs for a very, very long time. Which is also statistically impossible. If any species is reduced to a very small population over a long time period, some accident of nature or weather will cause their extinction. There is just no natural process that will produce a species with no genetic diversity."

Thor asked, "Then what's your explanation?"

Joe responded, "Well, isn't it obvious? The universe is filled with worlds sharing a similar biology and the cheetah is the product of an extraterrestrial breeding program. After all, it just happens to be the fastest animal on Earth. And an extremely unusual animal with no genetic diversity seems to me a glaring example of extraterrestrial visitation. There was an alien hunting party in Africa, and a couple of their pets escaped."

Frank Vitale spoke up next. "We know that cheetahs were kept by royalty, particularly in India. It seems much more plausible to suppose that some ancient Indian civilization had inbred the cheetah, rather than to presume that extraterrestrials had done it. Remember Occam's Razor."

"Well, I'm not an expert in genetics," Joe retorted, "but I don't think any reasonable length of time with any reasonable number of animals could have resulted in such a lack of diversity. I really think it would take cloning to get there. And we know that nobody on Earth has been cloning cheetahs, at least not until recently."

"Alright," said Thor, "What's you next example?"

"Twenty years ago," continued Joe, "a botanist discovered a small group of pine trees—Wollemi pines—in Australia, not far from Sydney. This pine tree had been, as far as anyone could tell from the fossil record, extinct for 50 million years. But here you have a small group—less than 40 trees—growing in a small valley in some Australian national forest. From the fallen logs, it seems that they've been there for some thousands of years. But it strains credibility that this ancient tree should be found in one dense stand, without any other known examples or relatives anywhere close by."

Thor said, "And your explanation for this one?"

Joe replied, "Obviously, a seed from a Wollemi pine growing on some other world was tracked in on somebody's foot or landing gear a few thousand years ago, and the trees have been growing there ever since."

Thor said, "I've read about those trees. They're rather a popular item at the florists these days, sort of like Norway pines. I think they're bisexual—doesn't that mean that there would have had to be two different trees from at least two seeds? I think that's really a stretch."

Frank said, "Yeah, I've heard of them, too. From what I've read they're related more closely to trees in Australia than anywhere else in the world. Most likely it's just one of those improbable events which, as probability predicts, will certainly happen from time to time. After all, the gingko was thought to be extinct until it was discovered that the Chinese had been cultivating it for centuries."

Joe replied, "Well, maybe if you looked into the genetic diversity of the gingkoes, you'd find some evidence that they came from a single or a very few plants. But the point is, the Wollemi were extinct on Earth, but not on some other planet."

"Still," Thor said, "discovering living examples of supposedly extinct animals is not particularly unusual. Don't forget the coelacanth they found off the coast of Madagascar. Ok, Joe, what's your third example?"

"Archeans," Joe said simply.

Thor thought for a minute and said, "Okay, I've read about Archea, too. They certainly are interesting and mysterious. Microbes halfway between bacterium and Eukaryotes. Very few species can be cultivated, and they seemed to contain wild diversity in their genetic makeup. Many widely differing types, some so strange they make the difference between bacteria and green plants seem small. But what is the evidence that they are extraterrestrial? There certainly are a lot of strange life forms on this planet which can't be cultivated. But, especially given their wide distribution in soils all over the world, there doesn't seem to be any particular evidence that they didn't originate here on Earth."

Joe said somewhat irritably, "Well, okay, I may not be an expert on Archea. But assuming, based on the cheetah and the Australian pines, that the universe is filled with creatures that are not too distant genetically from us, and that they've been visiting Earth on rare occasions for a very long time, what would be more natural than to find classes of microorganisms that aren't particularly well suited to Earth environments, but would nevertheless become well established, probably growing slowly as the Archea do. They also seem to like extreme environments, such as hot springs and hydrovents."

THE discussion trailed off from there. There was so much we just did not know about our universe. I was reminded of an article that I had read recently that talked about the failure of the most recent experiments intended to find the dark energy and dark matter in the universe. It was looking more and more like the proponents of modern physics really should fall on their swords. At least it seemed clear that there were too many open questions in physics to be smug about our understanding of anything. Of course,

At least it seemed clear that there were too many open questions in physics to be smug about our understanding of anything.

despite the efforts of many modern physicists and cosmologists, the most important questions from a philosophical perspective—where the universe came from and where it is going—were not finding satisfying answers in the latest theories. Even evolution, which forms such a solid foundation for biological classification and for describing the relationships between species, needs to be rewritten every 50 years or so to take into account the contrary data that are continually being collected. This time around it appeared that the rewrite would end in failure. It wasn't that some new theory had come along to replace evolution. It was just that it didn't do a very good job of explaining all of the genetic information that was rapidly becoming available.

But, even so, technological progress was in most respects barreling ahead. The venerable desktop computer still hadn't arrived at stable simplicity, which everyone sought, and which everyone abandoned because the new features always seemed to be needed to get the maximum potential out of personal computers. Biotechnology and nanotechnology were hotbeds of research and entrepreneurship. But even with all of that, most of the economies of the world were stuck in a sort of malaise. In Europe and Japan, this could clearly be blamed on demographics, which showed alarming population decline. But, as is usually the case in human affairs, it didn't mean that the solutions were clear in everyone's mind. Socialism as an economic theory was as good as dead, but the world view and philosophy behind socialism and communism were still major influences in public thought. But it was hard to argue that there hadn't been some loss of faith in man's ability to know and to control all things. A new "great awakening" in the religious sphere did seem to be underway in the United States, which tended to emphasize the idea that perhaps man was not the measure of all things.

I came out of my reverie and looked around at the cafeteria full of AM&M engineers and technicians. I thought of all that I had seen and heard during my months working on this project. Whether or not low-cost space transportation would finally open up a new frontier and contribute in a positive way to the philosophy of this age, I really couldn't say.

CHAPTER 22
Halfway to Everywhere

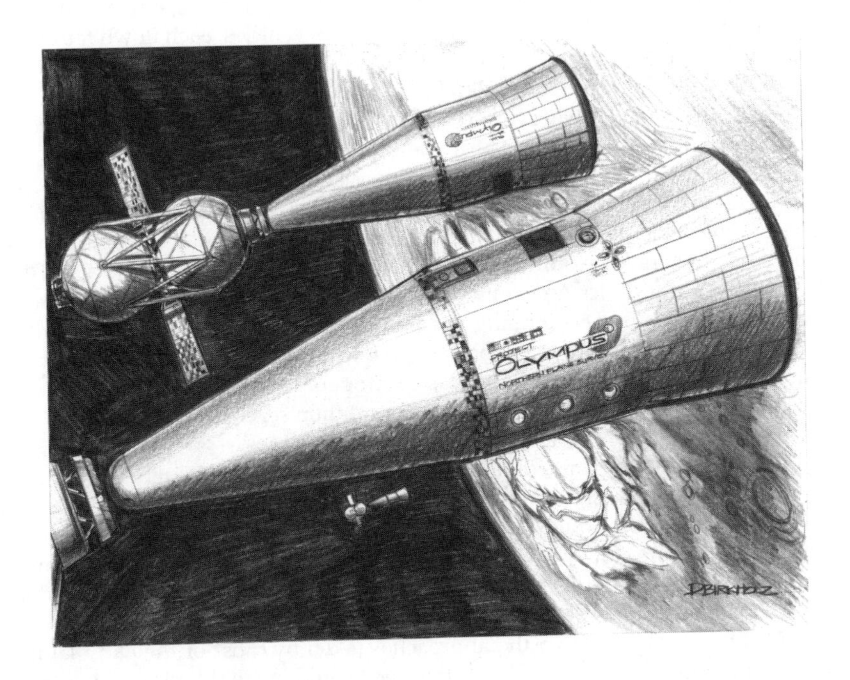

AFTER lunch, I headed up to Fred's office. On the way, I mulled over the conversations that I had had with John Forsyth about the potential of the DH-1 and his strategy of pushing it as far as it could go—literally. I kept hearing that low Earth orbit was half the battle, and Forsyth was hoping that by winning half the battle AM&M would emerge with the weapons needed to win the entire battle, which to Forsyth meant opening the entire solar system to human activity. For all of the progress being made in getting the DH-1 closer to its first orbital test flight, for all of the confidence he had that the engineers were taking a really solid approach, Forsyth was savvy enough to know that merely building a reusable space transport that didn't cost hundreds of millions for every flight was not enough. There had to be a real business at the end of it. Projections showed that AM&M would begin to flounder in approximately seven years if the market didn't really "take off" by that time. To survive to that point, they needed to sell an average of eight vehicles per year during those seven years. Seven years at eight vehicles per year would be 56 vehicles. Forsyth felt reasonably confident that they could sell that many in that time period, whether or not anybody had any profitable use for them. He believed there were enough national prestige, military, international cooperation, and other high visibility missions and programs such that—if the DH-1 were successful in doing what it was designed to do—there would be sufficient demand to keep the production line open for those first seven years or so.

But for the game plan to be successful by Forsyth's criteria, one or more new markets in space would have to emerge. Fifty-plus vehicles, each flown four times a year would make over 200 flights per year. Twelve times a year would be 600 flights. Flown once a week, it would come to over 2500 flights. Eventually, the buyers of DH-1s would have to find reasons to fly that often, or the market would begin to dry up.

Forsyth believed that success—in the sense of dramatically lowering the cost of space transportation—was inextricably linked to his ultimate goal: gaining access to the solar system. New markets for space transportation would have to arise, and in Forsyth's opinion those markets would be interplanetary. Thus, AM&M's space logistics group under Fred Clemens had a major role to play in Forsyth's strategy. By encouraging and jump-starting some deep-space activities, which might take 5, 10, or 15 years to come to fruition, Forsyth hoped to bring nearer the point in time when interplanetary operations would begin to drive the market for launch vehicles and space transportation services.

FORSYTH was determined to keep the company lean enough and squirrel away money enough to allow the company to survive on maintenance and operations support contracts, if sales tapered off because the demand for launch services grew more slowly than he hoped it would. But if new markets took more than 10 to12 years to develop, success might still come, but it probably would not be AM&M that reaped the benefits.

That was a far cry from the approaches taken by most of the early launch-vehicle entrepreneurs, who believed that they could make it by launching larger numbers of more-or-less conventional satellite payloads. There are only so many uses for Earth-orbiting satellites. I was beginning to see why Forsyth was looking so far away from home so early in the game. Significantly greater demand could only come from new markets in new—and bigger—territories. As I knocked on the door to Fred's office, I was anxious to hear about what he was working on for Mars and beyond.

Significantly greater demand could only come from new markets in new–and bigger–territories.

After exchanging pleasantries, I settled into a side chair as Fred moved over to the large white board that dominated one wall of his office. "Let me begin with an overview of the basic issues involved in interplanetary transportation," he began. "And let's start in Earth orbit—halfway to anywhere, as the saying goes. That's where we're going to begin, because our DH-1 orbital stage—or lunar, or interplanetary stage—will have to be refueled there before it can go any further. So we start with Earth orbital velocity of some 25,500 feet per second in a 100 nautical mile circular orbit. To leave, or escape, from Earth orbit, the speed of the vehicle has to be increased from 25,500 to 36,000 feet per second. At that velocity, the gravitational attraction of Earth won't slow the vehicle down to a complete stop until it reaches an infinite distance from Earth. For practical purposes, 'infinity' is a distance of one million miles. At that point, the Earth's gravitational attraction is effectively nil.

Now, if you want to go somewhere, you need some velocity above zero, right? So you need to leave Earth orbit with greater than escape velocity so that you have some velocity when you finally get beyond Earth's reach, so to speak. That velocity is called the hyperbolic excess velocity. To get to Mars requires about 9000 feet per second beyond Earth escape velocity. This 9000 feet per second added to the velocity of Earth with respect to the sun will put the vehicle on an elliptical, minimum-energy Hohmann transfer orbit that will extend out to the orbit of Mars. The time required for the outbound trip would then be about 259 days.

"OK, so the velocity increment from low Earth orbit to escape is about 10,500 feet per second. However, to head off to Mars, you don't actually need to add another 9000 feet per second. If you add roughly 1100 feet per second—so that you leave low Earth orbit at 37,100 feet per second—your velocity at an infinite, or at least very great, distance from Earth will be about 9000 feet per second." Fred stopped as I signaled a "time out."

"Wait a second," I protested. " That doesn't make any sense. Where does the extra velocity come from?"

"That's a concept that most people have trouble grasping the first time—and maybe the second or third time," he conceded. "But the explanation is actually quite simple.

"You've heard the analogy with the planets residing in gravity wells, right? With the sun at the bottom of a deep depression on a plane, and the planets at various heights up the slope leading out of the sun's well, each in its own smaller gravity well. To get out of a gravity well, such as the Earth's, you need a certain amount of energy corresponding to the depth of the gravity well. Our 36,000 feet per second at 100 nautical miles altitude corresponds to the energy necessary to escape Earth's gravity well from that orbit. Now kinetic energy is proportional to the square of the velocity, and so the energy necessary to escape from the gravity—or potential energy—well around Earth is proportional to 36,000 feet per second squared. So if a spacecraft is given a velocity greater than escape velocity, when it reaches a sufficient distance from Earth—again generally taken to be one million miles—it will have gained potential energy equal to the entire energy of the potential, or gravity, well and that energy will come out of the kinetic energy. So if we start from low Earth orbit at 37,100 feet per second, why do we end up with 9000 feet per second instead of 1100?"

I answered somewhat hesitantly, "Because we don't subtract velocity, but rather energy?"

"Bingo!" Fred exclaimed. "We square 37,100 and subtract the square of 36,000, and then take the square root of the result which comes out to 9000 feet per second. It's important to remember that when you're talking about orbital mechanics that as the speed of an object increases, its energy goes up as the square of the speed. A car going 100 miles per hour has four times the energy of a car going 50 miles an hour. For our spaceship, that 1100 feet per second of extra velocity means a great deal of additional energy, because it is added when the vehicle is already traveling at a high velocity.

"Therefore, our interplanetary ship will need a delta-V capability of about 10,500 feet per second to reach escape velocity and another 1100 to get to Mars. We've rounded that up to a total of 12,000, which when added to the orbital velocity of 25,500 feet per second results in a delta-V requirement of about 37,500 feet per second. And that is considerably less than the orbital stage is capable of achieving, which means that we can send a considerable payload to Mars, especially if we use another orbital stage for a boost like we will for the lunar flights."

"How considerable?" I asked. "Somewhat less than the 35,000 pounds to the moon, I suppose."

"Oh, about 130,000 pounds," Fred replied casually.

"You're kidding!" I exclaimed. "How can you send to Mars four times the payload you can send to the moon?"

"Well, don't forget that the moon flight includes returning the vehicle to Earth orbit. Of course, I'm also talking about returning the Mars vehicle, too. But that's not all cargo; actually, 90,000 pounds of it will be propellant for landing on Mars and departure for the return to Earth. But 130,000 pounds is enough for a small expedition to Mars."

Fred went on to explain that the "planetary stage" would be, like the lunar stage, a stretched cargo version of the basic orbital stage. It would be some 20 feet longer than the standard model, with landing gear added for use at Mars. The total empty weight would be 17,000 pounds, so that it could be launched into orbit, albeit empty. A major difference, though, would be the switch from a pair of RL60 engines to a single engine, with an engine bell extension to provide an impressive 468 seconds of *Isp*. The standard two-engine orbital stage had short engine bells, so they would not extend beyond the heat-shield moldline. For aerobraking at Mars or back at Earth, the extension would have to be removed and stowed. Even with the increased *Isp*, the initial launch of the planetary stage would probably require increasing the propellant load on the DH-1 first stage. However, each planetary stage would be boosted to orbit only once, and then stay in space, being based at the orbital refueling depot like the lunar stages.

After refueling in orbit, the weight of the planetary stage would be approximately 100,000 pounds, plus the 130,000-pound payload. Because most of that would be propellants, there would be an added pair of inflatable, spherical tanks mounted to the forward end of the planetary stage. Then, as with the lunar stage, the interplanetary ship would get a boost from another DH-1 orbital stage, in this case providing a delta-V of 4500 feet per second. The Mars-bound stage would then use the single RL60 with the nozzle extension and its full 82,000 pounds of propellant to bring the velocity up to 37,500 feet per second for Mars transfer.

The high-performance RL60 would also be configured to use propellants that are more "space storable" than the LOX and liquid hydrogen used for orbital and lunar operations because most of the propellants would have to be stored during the nine-month outward leg and the 13-month stay time at Mars. The propellants chosen were LOX and methane, which is what would be carried in the add-on tank module. Both are cryogenic, of course, but much easier to store for long periods of time than liquid hydrogen. This combination would provide a very respectable *Isp* of 360 seconds.

Before aerobraking at Mars, the planetary stage would take on 65,000 pounds of propellants from the tank module. The remaining 25,000 pounds of propellant and 10,000 pounds of supplies would be left in Mars orbit. The stage would then land on Mars, using about 25,000 pounds of propellant for a rocket-powered landing—requiring a delta-V of about 3000 feet per second—with 40,000 pounds left for the return trip to orbit. Some 20,000 pounds of supplies would be carried to provide for a 400-day stay on the red planet. At the conclusion of the surface stay, the planetary stage would return to Mars orbit using the remaining 40,000 pounds of propellant, delivering a dry weight of 20,000 pounds to orbit. After refueling with the 25,000 pounds of propellant waiting there and taking on the 10,000 pounds of supplies for the trip home, the planetary stage would boost into an Earth return transfer orbit and would use aerobraking on arrival to enter Earth orbit.

"WHAT about 'living off the land'?" I asked. " I know some people have looked at manufacturing propellants on Mars to reduce the amount that has to be shipped out there."

"Oh, yes," Fred responded, "refueling on Mars is a definite possibility. Robert Zubrin and the Mars Society have done a lot of work on the problem of making propellants on Mars, and there are a number of options. The Martian atmosphere consists mainly of CO_2, which could be broken down into liquid oxygen and liquid carbon monoxide. That would give you a specific impulse of about 260 seconds, if you use those propellants in the RL60. That, however, would only be sufficient to achieve orbit with a small payload, or for suborbital ballistic flights around Mars with a more substantial payload. But it wouldn't be enough for return to Earth. The liquid oxygen and methane that we are baselining is marginal for direct return to Earth from the surface, so you'd have to make several trips to stockpile fuel in orbit before return to Earth with a reasonable payload could be done."

"Why not make liquid hydrogen and LOX from the water on Mars," I asked. "That would certainly do, wouldn't it?"

"No question," said Fred. "Then you could easily go direct to Earth from the surface. Mars escape velocity is 18,500 feet per second at the Martian surface, compared to about 7,500 feet per second from low Martian orbit. Gravity losses and drag losses would be much lower, though, because of the lower gravity, thinner atmosphere, and the relatively high thrust to weight of the stage. The total velocity change needed for direct return to Earth comes out to about 19,500 feet per second. If you fully refueled the planetary stage with LOX and liquid hydrogen, you could place 30,000 pounds into Earth transfer orbit directly from the surface of Mars. That would allow a return payload of about 13,000 pounds—more than adequate for a crew of three to seven.

"Yes, we know that water is plentiful on Mars. There is certainly a lot of water ice at the poles, and many scientists suspect that subsurface water is also abundant. In fact, I've talked to more than one astrogeologist who would stake his reputation, if not his life, on being able to point to particular regions on Mars where a not-too-deep well could be sunk and water pumped to the surface. However, to do the liquid oxygen-liquid hydrogen thing, you're going to need lots of power for electrolysis. The best way to do that on Mars is to use a nuclear reactor."

Of course, space nuclear power systems had been the subject of several NASA and military space programs, and the Russians had actually built and flown a fair number of nuclear reactors in space, using rather inefficient thermoelectric conversion technology. Fred went on to tell me that AM&M had let a small contract to Bechtel for the design of a 100-kilowatt gas-cooled nuclear reactor. According to Bechtel, a suitable reactor could be developed with a core weight of 1000 pounds and at 2% burn-up could supply 100 kilowatts for about a century.

The cost would be in the neighborhood of $100 million. It would be combined with a turbine developed by Air Research, who had done studies on just such a combination of one of their gas turbines with a Bechtel-designed nuclear reactor in the 1980s. The design used gas dynamic foil bearings, so that once the shaft, on which were mounted the compressor and turbine, was brought up to speed there would be no physical contact between the shaft and the bearings, thus practically eliminating wear. Electrical power would be generated by mounting permanent magnets on the shaft itself. The working gas would be argon, which is available from the Martian atmosphere. It would be compressed by the compressor on the rotor, sent to the gas-cooled reactor, heated to a relatively low temperature, and expanded through the turbine. The intrinsic low environmental temperatures of Mars made the heat exchanger for this system relatively lightweight. In fact, the estimated weight for the entire system was less than 10,000 pounds, including fuel and the heat exchanger.

The biggest problem with any nuclear system is regulatory. Nuclear power, though it seemed poised to make a comeback, was still vulnerable to environmentalist opposition and consequent regulatory hurdles. However, Fred's plan for the Mars system involved procuring the reactor—unfueled—in the United States and buying the fuel elements in Russia. They would not be brought together until the components were set up on Mars. Although he knew there would be a firestorm of opposition to the very idea of a nuclear reactor on pristine Mars, he figured they should be able to work around the regulatory issues because there were as yet no regulations for the licensing of nuclear reactors on Mars. Certainly Fred and the others at AM&M were committed to the safest possible system. Nuclear fuel for nuclear reactors is hardly radioactive at all before the reactor has been energized. The enriched uranium fuel pellets for the gas-cooled reactor design that had been selected would be sealed within a ceramic coating, which would contain all of the radioactive daughter products. The reactor itself would be of simple, reliable, and robust design, requiring over its long lifetime only relatively infrequent servicing to adjust the control elements to maintain the reactivity at the desired level. Safety would be further ensured with the installation of a large tank of liquid carbon dioxide, which—because the thin Martian atmosphere would provide only minimal convective cooling—could be used to cool the reactor quickly if some catastrophic accident stopped the circulation of the working gas.

With such a power source available, using the resources of the Martian atmosphere and the abundant Martian water supply to manufacture propellants, transportation between Earth and Mars would actually be less difficult than between Earth and the moon. As Fred pointed out, landing on Mars is actually less energy intensive than landing on the Moon because the availability of an atmosphere makes aerobraking possible. Of course, it would take almost 9 months to reach Mars, compared with four days to the moon, so you would obviously need much larger stores of consumables for the crew.

... landing on Mars is actually less energy intensive than landing on the Moon because the availability of an atmosphere makes aerobraking possible.

If launch costs to low Earth orbit came down to $200 per pound, payload on Mars would cost $1200 per pound. The total cost for such a Mars expedition would include the cost of 300,000 pounds in Earth orbit, or $60 million, plus the cost of one specially outfitted planetary stage, which would be about $200 million. Then there would be the tank module, supplies, on-orbit refueling costs, and the use of an orbital stage for additional delta-V at the start of the trip. These, Fred estimated, would require another $50 million or so. Total cost of the mission would be about $300 million to send three to seven people to Mars for a round trip of some 950 days. The total number of orbital flights used for refueling and resupply would be in the neighborhood of 50, again helping to build the market for DH-1s and their launch services.

Even if launch costs were as high as $1000 per pound, the cost of the mission would be less than half a billion dollars. Fred told me that he and John Forsyth had been discussing the prospects of getting at least one nation to sponsor a Mars expedition like this, with the hope that maybe half a dozen nations would decide that it was worth participating for reasons of national prestige. Perhaps they could even be subtly encouraged to turn it into an actual race, with more than one expedition in the works.

IT was hard not to be somewhat skeptical about moving so quickly from an orbital DH-1 to lunar and Mars expeditions. On the other hand, Fred certainly made it seem plausible. I had come to realize that the folks here at AM&M, more than anything else, were without exception both practical and technically competent. As far fetched as some of these proposals seemed, it was hard to argue that they couldn't be done. "Anything else in the works for interplanetary missions?" I asked. "Or is Mars the limit at this point?"

"We've had one apparently serious inquiry about asteroid missions," Fred answered.

I knew that there had been a lot of interest in asteroids among space enthusiasts over the years. At least since Dandridge Cole wrote *Islands in Space* in the mid-1960s, the idea that there were gold mines in the sky—asteroids whose precious metal content was worth trillions of dollars—had surfaced from time to time. The early 1980s saw a lot of work on lunar and asteroidal resource utilization. Although many meteorites did exhibit relatively high concentrations of precious metals, especially in the platinum group, the more likely use for asteroidal resources was to provide more mundane materials such as nickel, iron, and water for use in Earth orbit.

"Which asteroids—the big ones, or the near-Earth types?"

"Actually, the Martian Trojans," came the unexpected reply.

"What, and where, are those?" I asked.

Fred proceeded to draw a sketch on his whiteboard. "They're a group of small bodies that occupy the Mars–Sun L4 and L5 points." He drew two large dots on an arc representing Mars' orbit, one 60 degrees ahead of and the other 60 degrees behind Mars. "You're familiar with the Earth–moon L5 and other Lagrangian points, I imagine."

I told Fred I knew about them as potential sites for space colonies, as popularized by Gerard O'Neill in the 1970s.

"Well, as you may know, for any two large bodies in space such as Mars and the sun, there are five Lagrangian or Lagrange points, three of which are unstable, and two that are stable, meaning that small objects placed at those points will tend to stay there. The L4 and L5 points are the stable ones. At this time, we know of

12 Martian Trojans. They got that title because the first one discovered—in the sun–Jupiter system—had been named Achilles, after the hero of Homer's Iliad, which was of course about the Trojan War. Anyway, these are asteroids that wandered in and became trapped at those points. Reaching them requires the same velocity that you need to get to Mars. If you fly by Mars for some aerobraking, you can reduce the velocity change required for rendezvous to somewhere between 500 and 1000 feet per second. That means that, if your vehicle leaves Earth orbit with a payload of 130,000 pounds, about 90% of that can be delivered to a Mars Trojan asteroid. Furthermore, if you fly by Mars again when you're ready to come home, you would need a delta-V of only 2000 to 3000 feet per second to return to Earth. On the downside, the trip times are long—years for a round trip—but the cost of transportation to the Martian Trojans need only be three to four times that to Earth orbit. And, again, you can use our same basic planetary stage."

"So, what about the near-Earth asteroids," I wondered. "Or the proposals I've read about for moving small asteroids into Earth orbit."

"There are some interesting candidates," Fred allowed. "Like 1982 DB, which every 10 years or so, has such a low Earth return velocity—less than 1000 feet per second—that a simple rope tow system on the asteroid could be used to launch small payloads back towards Earth. If you did that, and if the packages were small enough so that flying them through the atmosphere for aerobraking did not represent a hazard to people on Earth, the cost of delivering a pound to Earth orbit might be only a few dollars."

"What do you mean by a 'rope tow system'?" I asked quizzically.

"Like a ski lift," Fred said. Seeing I was still puzzled, he went on. "It would consist of a cable running around two pulleys, one of which would be driven by an electric motor. Instead of hauling skiers up a slope, it would take your payload and accelerate it along a straight course up to escape velocity and fling it off the end."

"That's a new one on me," I said. "But, I can see how it could work on a very small asteroid, with no gravitational pull to speak of."

"As far as bringing asteroids into Earth orbit," Fred continued, "I've always considered that to be pretty silly. I just can't see it being done, because the potential damage from a miscalculation would be so huge. Even Lloyd's of London couldn't insure against the global disaster that would result from even a small asteroid hitting the Earth, no matter how low the probability. However, I can see how large, million-pound class payloads could be sent back to Earth."

"Wouldn't you have the same problem?"

"If your payload carrier has something less than a million pounds of structure, you could design it so that it would break up into small, harmless fragments in the event of an aerobraking catastrophe." Fred elaborated. "Aerobraking large 'space barges' should be an economically viable—and safe—way to move large quantities of structural materials and volatiles like water and hydrocarbons into Earth orbit."

"Water, for example, is likely to be valued on the order of at least $100 per pound in Earth orbit for the a long time to come. In fact, that's what the company I mentioned earlier is proposing to do. They want to go out to one of the asteroids, mine enough nickel to build large lifting-body 'barges' using a proprietary nickel vapor deposition technique, and fill them with water. When the launch window is right, they use a rope tow system to start one on its way toward Earth. Of course, the barge would arrive at Earth with something greater than escape velocity—considerably more than 36,000 feet per second—but a large metal container filled with water will be able to absorb a lot of aerobraking heating. Even the most severe entry profile for, say, a bent-biconic vehicle weighing half a million pounds and carrying 20 million pounds of water, would hardly raise the temperature of the water at all. If it had 2000 square feet of surface area, and each square foot saw a heat load of 10,000 Btu, it would only raise the temperature of the water about 20 degrees. Then you have 20 million pounds of water in low Earth orbit with a market value of about $2 billion—a tidy sum, I'd say. And the empty barge would certainly make a good start on a space station or a small space habitat."

"That's an interesting concept," I conceded. "But it does sound like Mars' orbit pretty much marks the limit for DH-1 or derived vehicles' operations."

Fred whipped out his ancient 41-CX calculator, which was programmed to calculate orbital-transfer velocities for Earth, the moon, the major planets—anywhere within the solar system. "Well, you only need a hyperbolic excess velocity of 30,000 feet per second to reach Jupiter. The square root of 30,000 squared plus 36,000 squared is 47,000 feet per second in low Earth orbit. So by adding an additional 11,000 feet per second beyond escape velocity in low Earth orbit, you achieve an additional 30,000 feet per second at an infinite distance from Earth. Of course, a basic orbital stage refueled in low Earth orbit can achieve another 25,000 feet per second beyond orbital velocity for a total of about 50,000 feet per second—which would get the vehicle to Jupiter quite handily. Of course, that would be a very long trip in a very small ship. By the time somebody's ready to go to Jupiter, they'll probably want something more suitable. But the moon, Mars, the asteroids—that's a lot of ground to cover. And once we have the ability to get to low Earth orbit affordably and reliably, and we develop a support infrastructure there and on the moon and Mars, it won't be too hard to go anywhere."

FRED proceeded to summarize the few "simple tricks" that are the keys to moving around the solar system efficiently. "The first is, of course, refueling. Fred brought up a point that I hadn't thought about before, and that is the desirability of doing so after velocity increments of less than 15,000 feet per second or so. "The reason," he explained, "is that you can get so much more payload to the lower velocities, since your mass ratio can be much lower. Even so, you could profitably do it after increments of 25,000 feet per second or even 30,000 feet per second with chemical propulsion, depending on the mission. The second trick is aerobraking, which eliminates most of the propellants needed to reduce your velocity to match speed with your target planet—if it has an atmosphere.

"The third trick, often used with space probes, is to shape your trajectory so that you fly close by a convenient planet. By doing so you can bend or turn the path of your vehicle without any expenditure of propellants. You can also use a swing-by to gain or lose velocity. That one is probably the least obvious. However, you can do it because you typically approach a planet traveling either faster or slower than the planet travels in its own orbit around the sun. This velocity mismatch looks the same whether the spacecraft is going faster than the planet or slower when you get in close to the planet itself. For example, if you leave Earth orbit with about 20,700 feet per second over orbital velocity, your hyperbolic excess velocity will be about 29,000 feet per second. Following an elliptical orbit towards Jupiter, you will arrive at Jupiter's orbit going about 18,000 feet per second slower than Jupiter's orbital speed around the sun. But now you can use Jupiter to change your direction so that the 18,000 feet per second too slow is now 18,000 feet too fast, sending your craft on out to Saturn, or Uranus, or beyond. Or you can take the 18,000 feet per second and subtract it from your already too low velocity so that you to fall back nearly all of the way into the sun. Jupiter, the most massive planet, is the most useful for this type of maneuver, but Venus, Earth, Mars, or even the moon can be used, if you are patient enough to make several passes to gain or lose velocity.

"Another trick discovered by Dr. Edward Belbruno is that, if you take your time, it is always possible to approach a planet with zero hyperbolic excess velocity—with respect to the planet—simply by choosing your point of entry into the planetary sphere of influence. This means that the maximum velocity that would need to be removed by aerobraking or rocket thrusting would be only one minus the square root of two times the orbital velocity.

"Lastly, it is important to remember the advantage of adding velocity when your velocity is already high. That's what allows you, starting in low Earth orbit at 25,500 feet per second to turn an added 1100 feet per second into 9000 feet per second of hyperbolic excess velocity. Thus, if you want to speed up a spacecraft, or slow it down, it is best to do it closest to the planet when your velocity is high. For instance, if you want to rendezvous with one of the Martian moons Phobos or Deimos, you approach Mars at Martian escape velocity of about 18000 feet per second. Then all you need is a small burn near the top of the Martian atmosphere, or better yet some aerobraking, to put you in an elliptical orbit intersecting the orbit of Phobos or Deimos. Then you only need about 2000 feet per second to circularize and match the orbit of your target moon. Or, if large amounts of propellant were available at the L4 or L5 point in the Earth–moon system, a spacecraft could achieve near escape velocity and enter an orbit that would take it to one of those points for refueling. Then as its orbit brings it back around the Earth to perigee, it could fire its engines to increase its speed by another 10,000 feet per second and end up with a hyperbolic excess velocity of nearly 30,000 feet per second."

Thus, if you want to speed up a spacecraft, or slow it down, it is best to do it closest to the planet when your velocity is high.

Fred mentioned that when Father Scipio was here, he had told of an interesting concept that one of his monks was working on for the ultimate desert hermitage—a one-way trip out of the solar system. This young monk figured that if a man in his late 20s or early 30s followed the new life-prolonging, very low calorie diet, he would require only about 1½ pounds of supplies per day. Assuming a life expectancy of 100 years beyond his launch date, some 50,000 pounds of supplies would be required. A spaceship could be built with a rotating cabin section to provide 10% of Earth-normal gravity, connected to the nonrotating section by a magnetic fluid vacuum seal. Mounted to the nonrotating part would be a first-class astrometric telescope and a 100-meter radio telescope. The basic spacecraft would include a cone-shaped heat shield and a solar-heated hydrogen rocket. The approximate dry weight of spacecraft, supplies, and instruments would be 100,000 pounds.

Launched from low Earth orbit to a rendezvous with Venus, and using Earth–Venus exchanges, or flybys, the spacecraft could make it out to Jupiter in seven or eight years. At Jupiter, the velocity would be reduced to allow the spacecraft to fall down towards the sun to within 20 solar radii. As the spacecraft approached the sun, it would look like a huge reentry body with the sharp end of the cone pointed towards the sun. Deep in the gravitational field of the sun, the vehicle would fire its engine to get a 30,000 feet per second increase in velocity. The hydrogen could be heated to about 5000°F, just within the capabilities of high-temperature ceramics. With that temperature, you will get disassociation of the hot hydrogen, and the solar rocket would give an *Isp* of 1400 seconds, with the chamber pressure at a mere 15 psi. So close to the sun, 1 to 2 pounds of hydrogen per second could be heated with relative ease, and the engine would expend some 125,000 pounds of hydrogen during the 24 hours when the spacecraft was closest to the sun. The 30,000 feet per second, added deep in the sun's gravitational field, would propel the spacecraft out of the solar system at over 150,000 feet per second, or almost 10 astronomical units, or AU, per year. At distances between 100 to 1000 AU, measurements could be made to determine very accurately the distance to objects thousands of light-years away.

According to Fred, such a mission would probably have to wait until asteroidal water was available for a few dollars a pound in low Earth orbit to be affordable. But even at $200 per pound, such a mission could cost less than $100 million—less than a low-cost NASA planetary launch. Another way to do it, Fred suggested, would be to pick up 75,000 feet per second by diving into the cloud tops of Jupiter and using rocket thrust to add another 20,000 feet per second. Without being close enough to the sun to use a solar hydrogen rocket, or without using a heavy nuclear rocket, that 20,000 feet per second would require a mass ratio of more than 3.5 with chemical propulsion. With the solar rocket, 30,000 feet per second near the sun would require a mass ratio of just a little better than 2. Still, it might be easier to survive the radiation belts of Jupiter than to survive within 20 solar radii, even during solar minimum.

We hashed over a few more concepts that potential customers or other space enthusiasts had devised. Some of the scenarios that Fred outlined for me seemed very hypothetical, or at least something for the distant future—sending asteroidal water or ammonia to a crash landing in the dark craters of the lunar poles, for example. Nonetheless, I was beginning to get a sense for just how much would or could be happening in space within the next few years.

SHORTLY after my talk with Fred, I made another circuit with Forsyth and Tom to visit some of the close-to-home customer prospects. It was interesting to see how Forsyth used some of these farther-out concepts to stimulate interest from other potential buyers. For instance, the prospect of 20-million-pound lifting bodies aerobraking in low Earth orbit had a palpable effect on a group of Air Force officers that we met. It certainly made them look at the mission they had been seeking since the early 1960s—controlling the high ground of space—in a new light. If space transportation costs fell low enough to open up the solar system, it would surely be desirable to have someone take on the role of policing at least translunar space. I had even seen Forsyth pass out copies of Heinlein's *The Moon is a Harsh Mistress* to anyone who was not familiar with it. Forsyth provided the impetus to get the Air Force to begin seriously studying the implications of low-cost space access. Small study contracts were let for things like a geosynchronous Air Force base and the radar, lidar, and communications systems that would be necessary for traffic control and surveillance of the inner solar system.

Forsyth was more confident than ever that the DH-1 would make manned access to space commonplace and lower the cost of space transportation to the extent necessary to create a paradigm shift, which would open up countless new and as yet undreamed of human opportunities. Overall, though, it was his intent to avoid overblown claims, and he tended to downplay talk of futuristic lunar or Mars missions. He kept AM&M out of the Sunday supplements, but there was a steady stream of relatively low-key articles in *Aviation Week*, *Space News*, *Aerospace America*, and various other technical magazines.

Forsyth and Tom stayed the course of building an appreciation and understanding of how the system would function and the technology that it would utilize.

In an aerospace culture where the repercussions that something like the DH-1 would generate were so out of step with the attitudes and experiences of the big industry players, no real apprehension of the impending change was yet evident, and so institutional resistance was virtually nonexistent. NASA had just completed its preliminary design review for the Mach 6 hypersonic booster, but already it was becoming apparent that the weight and performance targets were not likely to be met, and not a few managers were beginning to look around for something that would distract attention from the looming cancellation of the program. It was beginning to look like the next year or so would be an ideal time frame for the start of the prototype flight-test program.

CHAPTER 23
First Stage, First Flight

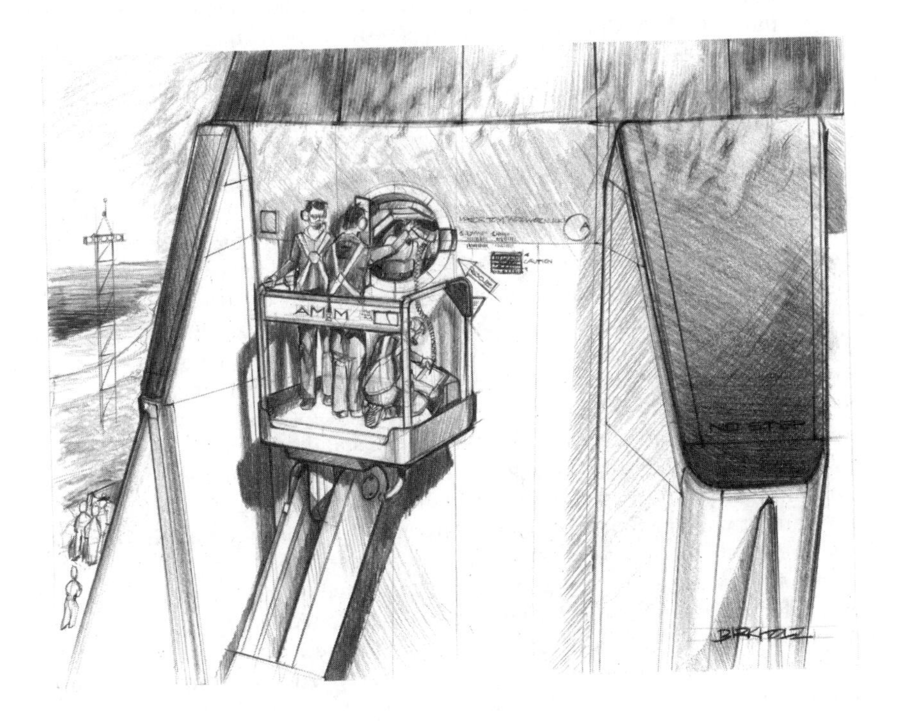

IT had been just about two years since major funding for the DH-1 had been committed, and the first major milestone was rapidly approaching—first flight of the first stage prototype. Three first-stage prototypes had been built or were under construction; one was earmarked for this first flight test, and a second was currently at the Cape undergoing checkout and test in preparation for the first orbital launch, now less than 6 months away. The third prototype was essentially a backup and would not be completed unless it was needed.

A government agency, even one as supportive of commercial spaceflight endeavors as the FAA's AST, was naturally going to be a little reluctant to approve the first private manned orbital flight, with the attendant possibility of loss of life. Tremendous public scrutiny had always been the result of any astronaut fatalities since the dawn of the Space Age. This first-stage test flight, while technically a flight into space, was following a trail already blazed by the X PRIZE contenders and the suborbital vehicles that followed. From a regulatory point of view, it was more like a first flight of a particularly exotic experimental airplane, as opposed to a manned reusable launch vehicle, and the permit had been issued with only a few minor delays. It helped that the first stage was capable of engine-out performance that would allow it to complete the mission if one engine had to be shut down. Even with two engines out, the vehicle could remain airborne until propellants had been exhausted or dumped and return for a soft landing. Part of the reason for flying the first stage with a dummy orbital stage was to lower the risks for the first orbital shot, of course, but also—from the regulatory perspective—it was intended to exercise the permitting process gradually as the first orbital flight loomed.

As has always been the case with multistage rockets, there are problems associated with flying the first stage without the upper stages on it because the center of gravity is not where it is supposed to be. The flight control system of a stage is designed to ensure control throughout its flight regime, and for a first stage that means when the rest of the stages are on top. But if the center of gravity moves closer to the engines, while the center of pressure does not, the vehicle can become so unstable that it is simply not controllable with the thrust vector control capability that is designed into the stage. One of the arguments for all-up testing of a complete vehicle, as opposed to incremental testing, is that sometimes the incremental test creates problems that aren't present in an all-up launch, and so you spend time and effort solving problems that don't really need to be solved. In the case of the DH-1 prototype, the need to maintain stability margins made it necessary to add a dummy orbital stage. This was a simple aluminum cone, almost identical in shape to the real orbital stage. For the test flights, it would be filled with water, giving it a total weight of approximately 60,000 pounds. The plan was to fly the combination up to 200,000 feet on a standard ascent profile, eject the dummy stage with springs, and return the first stage to the launch site. The site selected for this test flight was White Sands Missile Range, New Mexico. Although the orbital stage could be delivered from orbit to any location in the world, the first stage could not and would thus need to be shipped to customers' particular launch sites. The design of the first stage allowed it to be broken down into four parts so that it could be shipped by rail, sea, or air. This one had come in by rail.

Of course, you have to be careful when dropping a 60,000-pound dead weight from 200,000 feet, even if you do it at some place like White Sands. The damage potential of something that big falling that far is just too great, especially if it might come down uncomfortably close to the launch site because of the vertical flight profile. The test planners were counting on predictable winds to be sure that the dummy drifted downrange and away from the launch complex. To prevent the dummy stage from drifting beyond the confines of the range, though, it would not be equipped with a parachute, only a drogue to provide some stability and drag. To further reduce the possibility of damage on impact, the water would be drained after separation, using shaped charges to cut several openings in the bottom of the tank. To make the water flow out rapidly, some sort of pressurization scheme would be necessary. With instrumentation, the drogue, and now a pressurization system, the dummy stage was becoming more expensive and bothersome than was desirable for a one-use test article. Fortunately, one of Tom's engineers came up with the bright idea of filling the tank with carbonated

Of course, you have to be careful when dropping a 60,000-pound dead weight from 200,000 feet, even if you do it at some place like White Sands.

water. Some quick computer modeling showed that a roughly constant 10 psi would be available to force the water out of the tank; as water flowed out, the pressure would start to drop, which would cause more carbon dioxide to come out of solution and keep the pressure more or less constant.

In keeping with AM&M's low-profile approach, only a small number of dignitaries from the Air Force, NASA, a few potential private customers, and a smaller number of foreign representatives had been invited. There was, however, a sizable contingent from AST. I was surprised to learn that Forsyth had not invited the press—in fact they had been banned. He did not want any possible accident during the test to show up on the nightly news, at least not right away.

Early on the appointed day, I arrived at the White Sands Missile Range visitor's gate, where I was met by Chris Tombal. He told me that everything was on schedule, and that the launch seemed certain to get off that afternoon. But "certainty" for first launches of a space vehicle can be elusive. For the shuttle's first flight, *Columbia* had required almost two years of preparation, 105 days of which were on the launchpad. The DH-1 first stage had been here for four months, undergoing various tests and checkouts, including numerous cold-flow and hot-fire tests of the engines. The past two months had been devoted to using the rocket landing system to make short hops up to about 5000 feet in altitude. The final preflight tests, conducted just days ago, involved hot firing the main engines. This flight test would include about 84 seconds of powered flight, then 40 seconds of low thrust coast, followed by about 100 seconds of engine-out coast up to apogee, followed by a rather benign reentry. The whole flight—takeoff to touchdown—was going to take less than 10 minutes.

But "certainty" for first launches of a space vehicle can be elusive. For the shuttle's first flight, *Columbia* had required almost two years of preparation …

I talked with some of the test engineers during the morning, and they assured me that the wind predictions looked good for the launch. It was important that the winds be gently downrange, so that the dummy upper stage would separate and drift away from the launch complex. A couple of small reinforced concrete bunkers were located to one side of a very small viewing stand, and the dignitaries had been put through a few quick drills to make sure that they could get inside the blast shelters in less than 60 seconds. This would leave plenty of margin for determining that the dummy stage drogue had deployed, that the water had drained from the tank, and that it was drifting downrange as intended. The launch site was positioned a mile and a half uprange of the viewing stands. The launch was scheduled for late afternoon, and by midday it was hot, although it was a dry heat. Plenty of cold bottled water and soft drinks were being consumed as we waited for the beginning of the final count.

The chief test pilot, Jake Hill, was driven out to the vehicle about an hour before launch time, and I rode with him. After a thorough walk-around, and a lengthy discussion with the ground crew, Jake climbed into the pilot's seat. The vehicle was simple enough that putting a man on it seemed to not a few observers to be a bit of a stunt. Perhaps it was, but with a view to the regulatory hurdles that still remained, Forsyth wanted to emphasize the fact that both stages were going to be manned, and so the DH-1 could be treated more like an airplane than a rocket when it came to permitting flights near populated areas. But to me, when you considered that the landing propellants included a margin of only 15 to 20 seconds for hovering and landing, having a skilled pilot onboard made a lot of sense.

FOR the first stage, as would be the case for the complete DH-1, there was no gantry or umbilical tower. The problem with having a tower next to your launch vehicle, as Chris explained, is that you might occasionally fly into it. All ground power and signal connections, as well as the fluid supply lines, were connected to the bottom of the vehicle. Even the orbital stage—in this case, the dummy—was supplied via the first stage. At T-30 minutes, I joined most of the support staff and drove the mile and a half back to the viewing stand next to the launch control center, located in one of the bunkers. Of course, the vehicle was basically entirely self-contained, and the pilot could check it out himself and initiate launch without outside help. The launch control center was there to serve as a communications center, gather data, show the dignitaries what was going on, and generally give the regulators and government officials a warm feeling that things were under control.

At T-15 minutes, we watched the video monitors as the canopy was sealed up, followed by a final quick walk-around by the crew chief to be sure that the ejection seat was armed and that all of the "remove before flight" tags had indeed been removed. Then he and his assistant drove back to the control center. When the count reached T-5 minutes, a klaxon sounded. Chris had told me that, because the vehicle's trajectory is straight up and straight down and the total drift of the dummy stage was expected to be less than 10 miles, only the missile range proper—40 by 40 miles square—had been cleared out for the flight. The additional, 40 mile by 40 mile extension area, which was sparsely inhabited by cattle and a few ranchers and could therefore be cleared out for very little extra expense, had only been put on alert.

There were no sounds now, save for the occasional comment from Jake or the control center, relayed to us over the loudspeaker.

Control: Will you verify your switchover to vehicle power?
Jake: Roger. System looks good.
Control: Thank you. One minute.
Jake: Got prime. Igniter ready light is on.
Control: 30 seconds.

At T-1 second, the automatic ignition sequencer opened the valves, and ignited all five engines. A second later, the vehicle flew off the four launch posts.

Jake: I got a good light . . . and up we go . . . .

With over half a million pounds of thrust, the vehicle climbed remarkably rapidly. Sitting only a mile and a half from the launch stand, it took 7 seconds for the roar of the engines to reach us. In that time, the vehicle had already climbed to 350 feet. Hydrogen peroxide and propane have enough carbon in the exhaust so that it was bright and yellow, and a solid contrail followed the vehicle as it climbed toward main engine shutdown at 100,000 feet, 84 seconds into the flight.

Control: 50,000. How's your track?
Jake: Rog. 50,000. I'm drifting a little to the left . . . Bringing it back . . . .
Control: 70,000. Looking good.

Suddenly I noticed, as did those around me, that the vehicle seemed to be coming towards us. We were all well aware that the vehicle was not supposed to be going anywhere but up, and therefore should not be moving too far away from the launch site. But the idea that it might begin to drift directly overhead had a few people glancing towards the blast shelters. The AM&M team members who were sitting with us quickly reassured everyone that it was only an optical illusion. As the vehicle flew basically straight up, the distance between us and the point of launch became less and less significant compared to the altitude of the vehicle, with the result that the vehicle appeared to be climbing towards the zenith directly over our heads. Even after being reassured, it was an unnerving experience.

Jake: Vernier ready light on . . . Main engine shutdown in 3, 2, 1 . . . and off. Vernier engines are . . . on.
Control: Right on the money.

I had brought along binoculars—10x power, gyro stabilized—and could see the four distinct contrails of the low-thrust Vernier engines as the vehicle continued to coast upward, slowing just perceptibly.

Control: Coming up on 180,000.
Jake: Roger. Arming separation system . . . got a green. Staging in 20 seconds . . . .
Control: We see it . . . looking good from here.
Jake: She came off a little cockeyed . . . OK, I've got good separation. Verniers off.

I could see the dummy second stage separating, followed by a momentary restart of the Vernier engines, which served to make the dummy upper stage and the vehicle separate more rapidly. The vehicles were now at over 40 miles in altitude, and even with the 10x power binoculars I was having trouble distinguishing details on them. I counted slowly to 20, when the engineers had told me the water was due to vent from the dummy. Suddenly it was engulfed in a cloud, which was only slowly stripped away from the vehicle by the thin atmosphere at 200,000 feet. The dummy stage and the first stage continued to separate, very slowly it seemed, through my binoculars. At 60 miles up and only a couple of miles downrange, both still seemed to be directly above us, although I could hear from the loudspeaker that the radar skin track on the dummy was giving a predicted impact of over 4 miles downrange. The dummy had a simple cold-gas attitude control system to keep it from tumbling as it fell back toward the atmosphere.

Jake: Quite a view up here . . . 300,000 . . . RCS is . . . 10 degrees roll . . . .
Control: You should be almost to the top. How's your track?
Jake: I'm on the ellipse. Vertical speed heading for zero . . . Looks like I topped out at 341,000.

Jake had joined that exclusive corps entitled to wear astronaut wings, having now flown above 50 miles altitude.

Control: OK, 325,000 on the way down.
Jake: Rog. Backin' her down . . . Getting a little drift to the right . . . .
Control: We've got the dummy coming down four plus miles downrange. Shouldn't get in your way.
Jake: Roger. Peroxide pressure is 832.

As the stage reentered the atmosphere, it generated no fiery contrail. Reentry from space—from orbital velocity or greater—is normally a hot, tricky business, but coming in at 3000 feet per second, it wasn't any worse than the heating on an SR-71, and there was no visible change to the vehicle through my binoculars. Chris told me that the stagnation temperature remained under about 800°, and, in fact, the pulse was so short and the temperature so low that only the thinnest, lightest-weight components of the vehicle were in any danger of overheating. To design the stage for reentry, then, the engineers had simply made sure that there were no unprotected light-gauge pieces of sheet metal or the like on the windward side of the vehicle.

Control: 150,000. Please verify drogue is armed.
Jake: Affirmative, drogue is armed . . . a little buffeting there . . . RCS is fading . . . .

At 120,000 feet, Jake deployed the first-stage's small drogue chute, which had four shroud lines that allowed him to control the attitude and to a limited extent, the direction of flight of the stage. The dummy stage had deployed its drogue moments earlier, and it and the first stage, with its higher density and the controllability afforded by its drogue system, were now well separated. The dummy was clearly moving downrange and away from us, and the first stage was heading back down toward the launch site. There seemed to me a notable decrease in tension among the spectators as it became clear that the spacecraft were not going to drop on our heads.

The stage grew in size until it was almost halfway down the sky. Chris told me that the landing engines would be coming on very shortly.

Jake: . . . alarm for 12,000 . . . Vernier ready light is on. Peroxide pressure is 837.
Control: How's your track? Winds are steady at 5 knots.
Jake: . . . right where I want to be . . . Verniers on in 3,2,1 . . . OK, here we go . . . .

There was an audible, collective sigh of relief around me as we watched the vernier landing engines ignite. The hydrogen-peroxide thrusters produced only steam and oxygen at about 1200°, and it was rather hard to see the jet of steam around the vehicle, as it began its powered descent.

Jake: . . . vertical speed zero . . . I'm off to the right a bit . . . peroxide low light just came on . . . and I'm down.

In 30 short seconds, the vehicle had come to a hover at about 30 feet over the landing pad, and then rapidly descended to what looked like a perfect landing, with 10 seconds of fuel onboard.

Control: Very nice, Jake! The posse will be there pronto!

The AM&M engineers all piled into whatever vehicles were available and headed out to spray Jake with champagne. The VIPs and guests were offered some as well. I walked over to John Forsyth and Tom and congratulated them heartily. Perhaps in modesty, perhaps because it was true, they emphasized that although this was an important step, it was still a small step. Not really much more impressive than the DC-X had been back in the early 1990s. And even the DC-X hadn't really broken any new ground. It was something that everybody "knew" you could do. But, there was still satisfaction and excitement in doing well that which everyone knew could be done. The next stage of flight testing was going to be a tad more exciting.

CHAPTER 24
Stop the Production Line!

AFTER the test flight at White Sands, I spent a week visiting my sister in Texas before heading back to AM&M's Florida facility. I wanted to spend some time with my sister's family and leave the work behind for a few days, so I only checked my e-mail once or twice. Thus, when I walked into Tom's office the following Monday, I was surprised to find him up to his neck in alligators. The failure to invite the press to the test flight had backfired, and there had been a number of disparaging editorials and articles, hinting darkly that AM&M was working for the CIA, or part of some new superweapon program, or was involved in some foreign conspiracy. Tom said that he didn't know where the last one had come from, but there was one foreign-born member of the Group of Seven. Forsyth was working full time on damage control. For Tom, there were two other problems that were much more serious and immediate.

Both of the safety engineers and half of the reentry aerodynamicists—meaning Chris Tombal— . . . had "pulled the stop cord" and called a Toyota-style halt to the program. They were quickly joined by the test pilots. Apparently, there were two safety-of-flight issues that were not being adequately addressed. The first one involved the deployable heat shield that was to cover the engines and the engine

Both of the safety engineers and half of the reentry aerodynamicists had "pulled the stop cord."

bay during reentry. The current design required deployment of a large ring that formed the circumference of an umbrella-like heat-shield blanket. After the cover was fully opened, it would be pulled tightly against the bottom of the main heat shield. There were several problems that had cropped up. The deployment mechanism was proving to be tricky, and modeling and testing had shown that the deployable heat shield would cause shock reflections and severe hot spots on other parts of the main heat shield. That would require substantial beefing up of more than half the heat shield, for a large weight penalty. Even then, it appeared that a burn-through might still occur in the outer layer of fabric in several places.

The second problem concerned the landing gear for the orbital stage. The "boiler plate" test vehicle had already made a series of successful landings under both pilot and automatic control after being dropped from a CH-47 Chinook helicopter. In these drops from 12,000 feet, the parawing and landing gear were deploying consistently, the pilots and controllers had had little difficulty maneuvering to the landing strip, and although some of them were rather on the hard side, all of the landings had been successful. However, the landing gear on the test vehicle were considerably beefier than the gear destined for the production vehicle, where weight was critical. Based on the measured loads from the flight tests, the production vehicle gear would not be able to take the shear loads imposed by a typical landing.

The orbital-stage main landing gear were attached at the relatively thick ring where the cone-shaped upper portion of the vehicle was joined to the base heat shield, with its 38-ft radius of curvature. Two 5-inch diameter tubes extended about 7 feet into the tank to house the retracted gear, and their ends within the tank were unsupported. The original thought had been to run them across the tank and attach them to the engine bay, to better take the bending loads. The problem with that, as was quickly determined, was that if the ends of the tubes were attached to the engine bay, then high local stresses would be produced during pressurization of the tank and more so during cryogenic cycling. In addition, the quarter-scale test tanks were indicating a fatigue life problem with the landing-gear attach points after only some 50–100 cycles. Fatigue was not a problem for the test vehicle, but it certainly would be for the production vehicles.

Tom's team had dealt with troublesome design problems before, but these two came up together at a critical time. The end of the first billion dollars was fast approaching, and John Forsyth, did not think he could go back to the Group of Seven for more money before a successful orbital test flight, especially if the DH-1 had major unresolved technical problems. On top of all this, the press brouhaha over the "secrecy" of the first-stage test flight was causing some consternation within the Group of Seven, in that it seemed to call into question Forsyth's whole strategy for managing public perception of the program in order to generate and maintain confidence in the company and the vehicle. An aura of staid conservatism was going to be essential in order for almost any of the potential customers to obtain financing from government agencies or the capital markets.

WHEN it came to the engineering, well, Tom, as they say, had "been there before." He immediately pulled together a team of engineers from the related disciplines. Their first step was to thoroughly describe the problems and put them up on the intranet so that they could garner input from every member of the company and the outside consultants. The second step was to sort through the responses and pick out what looked to be the two or three best fixes for each problem, and then pursue them in parallel. The landing-gear problem seemed at first to be straightforward, basically a structures issue. However, it rapidly became apparent that the solution was going to require redesigning the landing gear, modifying the vehicle/parawing flying qualities to reduce landing loads, and, as it turned out, quite a bit of new software in the flight control system and modifications to the orbital attitude control system.

The problem with the heat shield seemed more serious. Undoubtedly, if you threw enough weight at it, you could solve it. But Tom was down to about 2000 pounds of payload, the prototype vehicle already being some 2000 pounds over the nominal design weight. In addition, it threatened to reignite the debate over whether the vehicle should be entering base first or nose first. A base first reentry requires some means to protect the engines from reentry heating. A nose first entry, on the other hand, keeps the engines safely on the downwind side of the vehicle, where little or no heat protection would be required. The shuttle main engines needed only the protection afforded by the body flap and some Rene 41 wire mesh shielding. On the other hand, nose first reentry would require heat shielding on the entire outer surface of the vehicle, which was four times the area of the base, and greatly increase the weight of the thermal protection system. Any of the candidate heat-shield materials were going to pick up some moisture in the form of condensation and frost, which constituted an added weight penalty as the vehicle carried that moisture aloft. The more shielding there was on the tanks, the worse this problem would be. Furthermore, during ascent the maximum aerodynamic forces—experienced as the vehicle moved through a max Q of over 600 pounds per square foot—would make use of the flexible ceramic blankets impractical on the windward side, driving the design toward the still problem-plagued shuttle tile solution. To make it even worse, if the vehicle did not come straight in, but at a slight angle of attack to limit g forces and provide some cross-range capability, the tanks would be heated asymmetrically during reentry. This would generate design-limiting loads because of the thermal gradient across the tanks.

There were some other advantages to base first reentry that Tom just did not want to give up without a fight. One of these was that the large spherical radius placed the center of pressure at the center of curvature of the heat shield, so that as long as the vehicle remained properly oriented towards the airflow the vehicle would be neutrally stable, and relatively small thrusters could keep the vehicle under control during reentry. Also, flying a small angle of attack for g load and cross-range control simply moved the center of heating from the center of the heat shield to a point about 60% of the distance out from the geometric center of the heat shield, which was beneficial in that the most severe heating occurred away from the engine compartment.

The landing-gear problem seemed to call into question the basic balloon tank design that Tom had pursued. The problem with pressure-stabilized tanks is that it is difficult to put large concentrated loads into the tank without large deflections. But adding material to beef up the structure in almost any place tended to have a deleterious effect on the fatigue life of the structure. The design of the

DH-1 was intended to achieve long fatigue life and reduce weight penalties, by attempting to eliminate structural reinforcements and the inevitable stress concentrations that they caused, wherever possible. The shuttle external tank (ET) on the other hand, used isogrid-stiffened panels and could be easily beefed up to take the large loads from the solid rocket boosters (SRBs) and the orbiter that hung from it. In fact, the ET owed more than 15% of its weight to such reinforcement.

While the engineers tackled the technical problems, Forsyth was on the road almost continuously, taking along one or two of the technical staff for three or four days at a time. They talked to every newspaper and publication that was interested. He also arranged to sell another $200 million worth of his assets to make sure there was enough money to get through the orbital flight test, bringing his total commitment to $700 million. Two other members of the Group of Seven put in another $30 million each, keeping their proportionate shares the same and providing some additional margin of comfort to Tom and his staff. AM&M's "burn rate," now that the test program was in full swing on top of the ongoing design work for the production vehicle, was right around $40 million dollars a month.

MEANWHILE, back at the plant, the initial teams that Tom had assigned to each problem were expanded into somewhat larger "Tiger Teams," one for the landing-gear problem and the other for the heat-shield problem. The Tiger Teams continued to use the open forum on the company intranet to solicit ideas and feedback from the entire engineering staff. This arrangement was beginning to bear fruit. The weight target for the landing gear was 1.5% of the orbital-stage empty weight plus payload, which was at the low end for airplanes. That came to about 250 pounds total, which allowed 105 pounds for each of the main landing-gear deployment mechanisms, strut, and landing pad. Modeling of the prototype landing-gear when subjected to the landing loads as measured on the boiler-plate test vehicle were showing near-certain failure of the main gear.

Unfortunately, the problem couldn't be solved simply by beefing up the landing gear attach points. That would just increase the stiffness in that area, which would exacerbate the tank's susceptibility to stress cracking there that had already shown up in the subscale tank tests. Throwing enough weight at the problem and ignoring the fatigue issue for the prototype could indeed solve the immediate problem, but part of the purpose of the prototype test program was to fly a vehicle that had essentially solved all of the engineering challenges in principal if not in production-level detail, and to demonstrate that some reasonable payload could be achieved in the production vehicles.

Within a week of the initial status report on the landing-gear problem, suggestions had begun to flow in. These included everything from air bags stowed in the cargo compartment that would inflate to extend over the rear end of the vehicle to retrorockets that would null out all landing loads. The Tiger Team put the suggestions into one of three categories: "possible solution," "part of the solution," and "unlikely solution." They worked long hours on analysis, simulation, and testing of the most promising approaches and regularly posted their results and findings on the intranet to garner additional refinements, suggestion, and critiques.

Within three weeks, an outline of the solution began to take definite shape. To provide more support and transfer some of the load away from the attach points in the tank wall, tubular aluminum struts, overwrapped with graphite epoxy, were added between the inside ends of the landing-gear pods and the central engine bay module/thrust structure. The struts were fitted within sleeves that accommodated the thermal expansion and contraction between the outer wall of the hydrogen tank and the inner wall of the engine bay. In addition, the area where the landing-gear pods penetrated the tank wall-base ring would be shot-peened at cryogenic temperatures to pre-stress these critical joints.

The original landing gear had shown a propensity to trip over the landing skids, which were simply ablative Teflon® pads mounted to forged magnesium feet. This was remedied by adding a small, high-pressure helium system within the landing strut, which fed a ring of nozzles on the landing pads. The gas system was triggered by a probe when the pad was 6 inches (or some 15–25 milliseconds) from touchdown, producing a "gas pad" hovercraft-like effect for about half a second as the pad contacted the runway. This served to reduce the horizontal friction force at the moment of landing that generated the high shear loads experienced by the gear attach points.

To reduce landing loads further, lightweight gas-actuated retractors were added to the parawing shroud lines, which allowed the vehicle to literally climb the lines to significantly reduce the vertical descent rate just as landing took place. This technique had recently been perfected by the Army for air-dropped cargo delivery.

The final element involved modifying the nose-mounted hydrogen-peroxide-propane thrusters so that they could operate at sea-level conditions. This involved triggering a uniform detachment of the hypersonic flow in the nozzles by injecting a small amount of compressed air in a ring around the inside of each nozzle, giving them an effective expansion ratio of about 8. The hydrogen-peroxide-propane thrusters located in the cargo bay were also redesigned to produce a total thrust of 2000 pounds. So with the cargo bay thrusters thrusting upward, and the nose thrusters thrusting downward—combined with the action of the shroud retractors—the landing loads were reduced by two-thirds.

The heat-shield problem also ended up requiring a combination of approaches and took even longer because of the need for modifications to an engine test cell at Pratt & Whitney.

The orbital-stage engines could in theory be cooled simply by running a small amount of hydrogen—less than 1 pound per square foot of engine area—through the cooling passages in the bells during reentry. Then the only other thing that would be needed would be a small, simple heat shield to cover the gap between the sides of the engine bay and the engine bells. However, because the hydrogen would necessarily be exhausted through the injector, there was some concern that it might detonate in the chambers as the vehicle descended into the lower atmosphere, or that it might leak from the bells back into the engine bay and mix with air to form an explosive mixture there.

An inert gas, such as helium, might eliminate those problems, but that would require the addition of a separate subsystem containing liquid or supercritical helium and the associated plumbing. Although liquid helium is relatively cheap—at about $25 per pound—and less than 30 pounds would be needed per flight, it went against the grain for Tom to seriously consider using a commodity like liquid helium. Its availability was strictly limited, being found worldwide only in a few natural gas wells, mainly in the United States. For most of the consumables used in the vehicle, such as the liquid hydrogen fuel, and even most of the hardware components, large numbers of vehicles and high flight rates would tend to bring prices down, but for liquid helium, it was more likely to drive it up.

But the major problem with the helium approach was that there simply wasn't enough time to develop and test such a system, in combination with a workable heat shield cover for the gap between the bells and the engine bay. Such a cover would have to accommodate the 3-degree gimbaling of the engines for thrust vector control, which meant that the bells could move up to 4 inches from their nominal positions. That would require that the heat-shield material around the bell be able to stretch some 4 inches in one direction and compress 4 inches in the other direction, and then return to a suitable position for reentry. Thinking about this problem led to the first part of the ultimate solution. A flexible, woven wire skirt made of Rene 41 was designed to fill the gap between the heat shield and the engine bells. The second part involved providing a water dump around the outside of engine bells during reentry.

Finally, the engineers decided to use, not an inert gas but liquid oxygen, to cool the engines during reentry. That would eliminate any explosion hazard, and the liquid oxygen would already be onboard. It would be supplied from the oxygen reserve tank, which was part of the orbital maneuvering system, or OMS. This system could hold up to 1000 pounds of hydrogen and oxygen in small tanks located within the main hydrogen and oxygen tanks, respectively. The small tanks were designed to withstand collapsing pressures of about 10 psi and were made with honeycomb walls of high stiffness and overlain with additional cryogenic insulation. When the main tanks were full, a vacuum was drawn on each of the reserve tanks, until an approximately 50% solid slush was formed. The OMS propellants could then be stored on orbit for up to two days, or allowing for some boil off, as much as one week.

On the downside, this did involve some significant redesign work on the engine system. To avoid major changes to the engines, the LH_2 prevalve was modified to incorporate a LOX cross-feed valve, allowing LOX to flow from the oxygen downcomer into the hydrogen supply line, and also a helium purge supplied from a tank in the payload bay. After the deorbit burn, the engines would be purged with helium gas to eliminate any residual hydrogen in the cooling passages or injector manifold. Then during reentry, some 300 pounds of oxygen would flow through the hydrogen sides of the engines, cooling the expansion bells and exiting through the injectors into the chambers. This flow of cooling oxygen out through the chambers and the bells would provide a further thermal shielding effect around the engine bay opening.

CHRIS Tombal, one of the instigators of the "stop cord incident," was quite pleased with the outcome, as I could tell during a review of the redesign effort. He concluded with a final point that would never have occurred to me. "There was one other concern that the controls people had; fortunately it turned out to be a non-issue."

"What was that?" I asked.

"Well, having 300 lb of oxygen up in the main tank during reentry would move the center of mass of the vehicle pretty far forward—toward the nose of the vehicle—and that would have a negative impact on stability and controllability.

But, as I pointed out, during reentry the max heating occurs well before max-Q, or max dynamic pressure. So most of the cooling LOX would be expended before max-Q, and the center of mass of the vehicle would be where we want it during the critical control phase at max-Q."

The heat-shield redesign added only 300 lb, taking into account the deletion of the deployable heat shield. Fixing the landing-gear problem added almost another 500 lb, including the beefed up landing gear, the shroud retractors, the modifications to the RCS system, and the additional landing propellants. Thus, the total "weight bill" was almost 800 pounds, which gave the prototype a nominal payload of 1200 pounds. It took almost six months to complete the redesign and reconfiguration of the orbital stage, including all of the necessary testing with the air-dropped landing test vehicle, until landing loads could be reliably brought within the capability of the redesigned landing gear. Fatigue life testing continued with the quarter-scale redesigned tank, and the stress levels were lower. Over 200 cryogenic cycles had been achieved without cracking, and it seemed likely that the design goal of 1000 cycles for the basic hull could be realized. Even 200 or 300 cycles would be adequate for the first production run of vehicles. One advantage of these problems cropping up together was that they could be worked in parallel. While the gear and tank redesign was being implemented, the propulsion engineers worked closely with Pratt & Whitney to hammer out and validate the modifications to the engine feed system.

DURING the six-month delay, John Forsyth's focus was to educate and reassure the industry, government agencies, and the public that AM&M was still on the right track, but he concentrated on those organizations that he considered to be the DH-1 launch customers. None of them—they were all government agencies at this point—were yet willing to go on record, but they seemed quite interested in getting in line for the first year's production. Forsyth was reasonably sure that he could turn this interest into firm orders, once a successful test program had been completed. There were eight of these tentative customers, and their identities were a closely held secret; Forsyth wanted to prevent any bureaucratic or competitive sidetracking of their plans before the vehicle was ready for production. He was more than happy to keep things quiet, as the potential customers worked to build coalitions within their agencies and lay the groundwork for commitment of the funds to buy the vehicles. The target price was still $250 million per vehicle, with an additional $100 million for related support facilities. The facilities portion would be subcontracted to Bechtel Corporation, with a 20% holdback for AM&M's contribution. To date, AM&M had hired Bechtel to do over $100 million worth of facilities work at Cape Canaveral, and Bechtel in turn had invested about $50 million of their own money in the program.

The target price was still $250 million per vehicle, with an additional $100 million for related support facilities.

The U.S. launch customers were the Air Force, for two vehicles, and NASA for one vehicle. They would be purchased from AM&M by a separate company, which would outfit them according to each agency's requirements and specifications, and then be sold with all of the requisite documentation as well as operations and maintenance support. The markup was expected to be about 50%.

The remaining five ships would be going to U.S. allies: two for ESA, two for Japan, and one for Australia.

For the foreign customers, Forsyth had a team working on the initial arrangements for financing, and AM&M was going to guarantee that the vehicles would perform at least 10 flights. The same companies that finance most of the world's airline fleet were actually quite interested in providing the financing for the first batch of DH-1s, sensing a lucrative new market for their services. Even with financing over a relatively short five-year term, in combination with an insurance policy that would pay off the remaining balance of the loan should the vehicle be lost after the first 10 flights, the total cost of ownership was expected to be approximately $100 million per year. It was abundantly clear that at low flight rates of two or three per year, the actual cost—on a per-pound-to-orbit basis—was not going to be a whole lot less than the shuttle. But the cost per flight was, by all indications, going to be less than one-tenth as much. With the advent of the DH-1, the incremental cost of getting into the manned spaceflight league—a mere $100 million dollars a year—would be dwarfed by NASA's budget for manned space operations, which, depending on how you counted it, totaled somewhere between $7 and 12 billion a year.

In the case of NASA and the Air Force, their chief interest was in using the vehicles as test beds for their own ongoing programs. The other customers were not particularly interested in dramatically lowering the cost of space transportation; they mainly wanted to "join the club" and become players in the manned spaceflight arena and gain the prestige that went with it. This was all fine with Forsyth, whose game plan called for getting a reusable launch vehicle into long-term production and selling as many airframes as possible. He had no doubt that, in time, as availability improved, payload lead times dropped, and even the cost per launch fell, the price per pound to orbit would start to plummet. But the first step was to find a way to sustain the production of a vehicle that was capable of achieving low-cost space transportation. The customer list for the second and third years' worth of production, although not as clear, could already be seen in outline: a new vehicle each for Korea, Taiwan, and Brazil; a second or even a third vehicle to Japan; a small squadron for the U.S. Air Force, and another test vehicle for NASA. Perhaps by that point, even France would decide that they needed their own, non-ESA vehicle—a space "flag carrier"—if they didn't want to be left out of a growing international stampede into manned spaceflight.

By the third year, Forsyth was counting on strong international pressure to make the vehicles available to any country that wasn't actually involved in hostilities or harboring terrorists. The technology incorporated in the vehicle was not particularly advanced, and it was—at least in theory—not beyond the capability of most European countries, China, or Russia. It would certainly be an expensive missile. At a cost of $250 million and a payload or throw-weight of 5000 lb, you could do a lot better by strapping together a few obsolete Korean missiles. As 9/11 had shown, there are few things more dangerous than fully fueled passenger jets, and yet those are routinely sold to practically every country in the world.

Forsyth had also formed a small task force that was working closely with the State Department and the National Security Council. Together they came up with a number of ideas for limiting the potential for misuse of a DH-1: things like always-active beacons that would continually broadcast the GPS-determined position of each vehicle to orbiting defense satellites, or special hollow microspheres that could be embedded in the orbital stage thermal protection systems. These would be highly reflective in certain microwave bands, thus enhancing the tracking signature. Indeed, the unique trajectory of the vehicle gave it an unmistakable Doppler signature in all optical bands that would make the launches easy to differentiate from other types of space or missile launches. Other systems were borrowed from the international regulatory agencies responsible for tracking commercial airliners for financial security reasons; indeed, most of the vehicles would be leased or financed in a similar fashion. Forsyth hoped that they would be able devise sufficiently reliable and effective systems and procedures to allow countries such as Pakistan, India, Israel, South Africa, various South American nations, and even Russia and China to buy vehicles without putting any nation at risk. It seemed likely that most countries would be willing to put up with quite a few restrictions, if they were left nominally in control of their own manned spaceflight capability. Forsyth had also gotten the attention and support of the Department of Commerce and several prominent members of Congress. They were quite interested in the potential for a new multibillion dollar export market for the DH-1.

They were quite interested in the potential for a new multibillion dollar export market for the DH-1.

To heighten international awareness of the vehicle, John Forsyth had commissioned a contest for the composition of a launch symphony in 35 countries or cultural regions around the globe that represented potential customers. The prize would be $50,000 for a composition lasting approximately 17 minutes, covering the time from 10 minutes before takeoff to orbit insertion. The idea had been a suggestion of a marketing consultant Forsyth had hired. The objective was to arouse national pride with special theme music for each nation's launchers. The music would be designed to underscore the dramatic tension and mounting excitement before the launch, replacing the countdown at T-10 minutes and getting people emotionally involved in the flight. The compositions were also to incorporate the actual sound of the launch as an instrumental component, just like the cannons in the *1812 Overture*. Each contestant was provided with a high-quality recording made at one-and-a-half and at three miles away from the first-stage flight test that had recently been conducted at White Sands, as well as a simulation of the sounds expected in the payload bay during ascent.

The existence of a national theme song could also cross over into popular music, raising the overall awareness of the vehicle and creating an expectation among the populations of the various countries that, in terms of national pride and taking their place among the great nations of the world, $100 million a year out of their national budgets would not be ill spent on such a highly visible means of national achievement. After listening to a couple of the early submissions that Forsyth gave me, I thought the consultant might be on to something. It certainly set a precedent for theme music for others who were pursuing various uses of the DH-1, from the transpolar lunar expedition to the various space tourism promoters. Hearing the theme music could serve to remind people of the excitement that they had felt the first time they watched a launch—on television, or even more so if they had been there in person.

But these things were still off in the future. The next step was a successful orbital flight.

Chapter 25
Earth Below Us

AFTER slogging through the six-month delay caused by the need to redesign the landing gear and the engine thermal protection system, the day of the first orbital flight finally arrived. Long ago Tom and his team had concluded that, although the vehicle could be recovered from any point along the ascent trajectory, the lowest risk option would be a direct flight all of the way to orbit, rather than a suborbital flight. Therefore, the first test flight of the prototype orbital stage would be the real thing. The flight would last for four orbits, to allow enough time to check out the vehicle and its subsystems before reentry. Paul Reston had explained to me how, by launching in a direction not exactly due east, you can make your ground track go over the launch site after any particular numbers of orbits at a cost of only a few feet per second of velocity. The idea that you need wings for cross range to enable once-around abort capability has always been somewhat misleading. Greater cross range would indeed provide more landing opportunities, but it's not necessary if you simply want to get back to your launch site after spending less than a day on orbit.

The vehicle had been on the pad for almost 30 days, and after several scrubs because of weather and a few hardware glitches, everything came together. It was a beautiful Florida morning, and although the AM&M engineers were a bit on the tense side, a party atmosphere prevailed. For this launch, the press had been invited, and there were reporters, newscasters, and writers from all over the world. There were also of course dozens of representatives from potential customers, and everyone was provided with food and drink. A small orchestra was on hand to play the launch theme, while over a loudspeaker came the occasional update from launch control. The more technically oriented guests stayed close to the control center so that they could watch the video monitors, hear the radio traffic between the test control team and the ground crew, and know exactly what was happening and when.

At T-1 hour, pilot Erica Phillips entered the vehicle, closed the hatch from the inside, and strapped herself into her ejection seat. Jake Hill did the same in the first stage. At T-30 minutes, a siren sounded, and cryogenic fueling through the base of the vehicle commenced. Tom had insisted on rapid fueling followed by launch within an hour, in order to hold down the risks associated with ice buildup on the vehicle. To keep condensation and ice buildup on the uninsulated liquid-oxygen tank to a bare minimum, the entire vehicle was covered in the lightweight Mylar balloon, which was cut with a hot wire and ripped away at T-1 minute. Fueling was complete at T-20 minutes. At T-10 minutes, the orchestra struck up the launch symphony, and the dramatic tension grew until at T-2 seconds the main engines ignited and built up thrust. Then, right on schedule, the DH-1 began rapidly climbing up the sky.

Everyone had been warned that the vehicle would appear to fly over—or rather, hang above—the grandstand. Even with my experience from the first-stage flight test, it was still hard to ignore the overwhelming sense that the vehicle was climbing right over my head. The hydrogen peroxide-propane combustion products of the first-stage engines left a clean, white contrail, which seemed to stretch from the launchpad directly to the zenith. After main-engine shutdown, there were 45 seconds of powered coast with the vernier engines, followed by stage separation. There was a just barely visible contrail that allowed you to follow the orbital stage with the naked eye as it moved eastward towards the horizon.

The orbital stage was, in theory at least, visible all of the way to the moment of injection into orbit. It was certainly in radio line of sight from the control center until orbital injection, and a direct microwave telemetry link ensured that all of the data would be recovered, even if the vehicle was not. All of the flight-test data, including camera feeds, were also stored onboard on high-speed data recorders. After an enthusiastic explosion of applause and cheering following the announcement that the vehicle had achieved orbit, all eyes turned to the returning first stage, which was now maneuvering under the power of its four hydrogen peroxide landing engines toward the landing area.

This was a one-mile-diameter circular area, centered on a 300-foot-diameter concrete pad with a central 100-foot-diameter area paved with cast-iron hexagons. These provided a smooth surface for the landing gear of the first stage, but more importantly, a surface that could withstand the direct thrust of the landing rockets. The cast-iron pavers could endure the direct impingement of the landing rockets for about 30 seconds without danger of spalling, thus minimizing the chance that loose chips of concrete or rock, accelerated by the jets, would damage the vehicle. The concrete pad contained a high percentage of steel reinforcing wire, which also served to minimize spalling. On this landing, Jake missed the iron pad, and brought the stage down on the reinforced concrete pad. There was no apparent damage to either the concrete or the vehicle.

AFTER more cheering, the crowd settled down to watch the progress of the orbital maiden voyage. The radio traffic was heavy on the engineer-speak as Erica performed checks and tests in concert with the engineering team on the ground, punctuated by the occasional comment on the view from up there. Attention was focused on the television monitors. Forsyth had arranged for a state-of-the-art HDTV downlink that allowed the guests to watch the Earth moving below the orbital stage almost as if they were onboard. During the six-hour flight, lunch was served, and all over the visitors' area people were engaged in lively discussions about what this all meant for the future. The AM&M sales team was exhausted by the time reentry began.

This was the phase that had the engineers most concerned. I can't say worried, because they had confidence in their engineering and that of their colleagues. Nevertheless, reentry is something that just can't be completely simulated or tested until you do it for real. Fortunately, this one came off without a hitch. We got excellent video feed from observers to the west, all of the way until the parawing was deployed.

Cheers erupted again when the vehicle, graceful under its colorful parawing, came gliding into view. The pilot had no trouble bringing her in for a very smooth landing. This time the applause and cheering were almost deafening—almost as good as a Packers playoff win at Lambeau Field. Now the engineers could really join in the party, which continued pretty much until dawn the next day.

Cheers erupted again when the vehicle, graceful under its colorful parawing, came gliding into view.

The flight got considerable play in the press, making the cover of *Aviation Week*, *Time*, and *Newsweek*, with a corresponding boost in public awareness. Jake and Erica appeared on several television and numerous radio talk shows, and a parade was put on by the city of Cocoa Beach, Florida. John Forsyth was pleased to see that for him and his marketing team doors were opening a little wider, and some that had been closed now opened for the first time. Attendance at his conferences and public appearances increased dramatically. However, the mainstream aerospace industry was not overly impressed. After all, nobody really doubted that you could build a manned orbital vehicle—with a tiny payload—for a billion dollars or so. The fact that it had succeeded was an accomplishment in its own right, but most in the business wrote it off as something akin to the round-the-world flight of the Rutan Voyager—interesting and laudable—but not the kind of thing that was going to change the world. Certainly the FAA's AST was happy to have a real live, manned launch vehicle that they could regulate. Overseas, however, the reaction was a little bit more enthusiastic, especially in countries that had aspirations to manned spaceflight and could now see the possibility of buying that capability for a heck of a lot less than the United States, Russia, or even China had paid for theirs.

THE discussions with the Japanese aerospace companies about coproduction agreements went into high gear. Forsyth was confident that the Japanese—with whom he had been patiently pursuing a gradual relationship-building process—would become major customers. Within the U.S. Air Force Space Command, there was a key group of officers who were definitely interested in getting their hands on a vehicle. Finally, after being effectively shut out of manned spaceflight for decades, they now saw a credible way for the Air Force to gain that capability, at a price they could afford. NASA, overall, was quite a bit more circumspect, but there was a definite movement within that agency to buy a vehicle. Not surprisingly, the major interest was centered in the science organizations, not the manned spaceflight centers. If they could purchase a reusable launch vehicle for hardly more than the largest airborne-telescope research aircraft cost, they didn't care where they got it.

Forsyth, who always believed in getting as much money as he could as early as he could in the financing of a new venture, called the Group of Seven together the very evening of the first orbital test flight. He laid out for them the achievements of the marketing campaign to date, the cash and in-kind contributions that various suppliers of major subsystems and components were willing to contribute, and the status of his negotiations with the Japanese. All told, these companies were ready to pledge $500 million—if Forsyth could raise another $2 billion from the Group, or from other investors.

Some of the Seven wanted to wait until the end of the flight-test program, when they would have a better idea of the real operating costs, the reusability of major items like the engines, turnaround times, and the final production-vehicle costs. However, Forsyth was well aware that the test program could still run into problems—maybe even the loss of a vehicle—and any such setback would make raising the funds necessary to get the vehicle into production that much more difficult. He was determined to raise the money right now, even if all of the facts were not yet in hand.

For this round, Forsyth could only contribute another $300 million, which constituted the bulk of his remaining net worth. That meant that $1.7 billion would have to come from the rest of the Group. Up to this point, the other six had invested $560 million, to Forsyth's $700 million. Another $1.7 billion would put a serious dent in their collective net worth, and for some it would put their level of commitment on par with Forsyth's, who was indeed putting nearly 100% of his fortune into AM&M. Fortunately, as Forsyth knew there would be in the wake of the first flight to orbit, there was a feeling of overall confidence that the project was on track and that it would be more than a technical success. Enthusiasm was running high that it was a major step towards lowering the cost of access to space. They were a little less sanguine concerning whether or not the investment would pay off for them personally. Drucker's Law—that the first firm into a new field inevitably loses it shirt—was never far from their minds.

Drucker's Law–that the first firm into a new field inevitably loses it shirt was never far from their minds.

In the end, Forsyth brought everyone around. He didn't really have to work too hard; after all, as he reminded them, their primary motivation in joining this venture had not been profits, but the opening of a new frontier. However, he had to make a major concession. Contrary to normal practice with young ventures like this one, the value of the existing stock was cut in half relative to the new shares, so that his own investment—representing about 65% of the money invested so far, and which should have amounted to a little over 30% of the company after the additional $2 billion infusion—was effectively reduced to about a 25% share. Forsyth had of course known that, as more capital was brought in, his share was going to fall quite a bit anyway. But it did mean that if there were too many reversals as the program proceeded towards production, the Group of Seven would be able to more easily oust him from control.

WHILE Forsyth was putting the production financing together, Tom and his team were busy analyzing the flight-test data and going over the vehicle stages with fine-tooth combs, gathering additional data and checking carefully for any damage. The heat shield had been stripped off the vehicle immediately upon landing and then sealed in plastic bags so that the amount of water loss in each section could be analyzed. Underneath the thermal protection shielding and on all exposed surfaces, the orbital stage had been coated with paint containing temperature-sensitive microparticles, which indicated the maximum temperature reached on the surface of the aluminum structure by changing color. Microsensors powered by heat flow had also been attached on 3-inch centers beneath and within the heat shield. After the flight, these were scanned with a special probe, using smart card technology, to provide a temperature time history for exit and reentry heating. So it became apparent shortly after landing that there had been some excess heating along the side of the vehicle closest to the stagnation point on the heat shield.

The center of mass of the orbital stage was offset very slightly from the vehicle centerline, so that during reentry a very small amount of lift—equal to somewhat less than 10% of the drag experienced by the stage—was created. This slight amount of lift provided some capability to correct for upper atmospheric flight-path perturbations, but, more importantly, it limited the maximum *g* loading during the reentry to just over 5 *g*. One consequence of the offset, however, was that the heat load on the heat shield was not completely symmetric. Maximum heating occurred at a point closer to the windward side of the vehicle. It was here that the aluminum substructure had seen temperatures between 400 and 470°F. This is not so high as to seriously degrade the properties of the aluminum, but it was a concern, especially because another, slightly different reentry profile might produce even higher temperatures. This problem was addressed quite simply by thickening the heat shield and increasing the water loading of the hydrate filler material around the overheated area. The amount of additional water was calculated to produce a cooler flow coming around the side of the vehicle, with the beefed-up heat shield serving as additional insurance. Total additional weight was held to 75 pounds.

The ablatively cooled engines in the first stage saw throat diameter growth of only about 50 mils, and it looked like the design life of 10 flights for the prototype was going to be achievable. A specially devised computerized axial tomography unit allowed 100% inspection of these engines without removing them from the vehicle. The orbital stage's RL60 engines were pulled and shipped back to Pratt & Whitney for a complete teardown and inspection. A new heat shield and new engines were then mated to the vehicle, making it ready to fly again in 30 days.

After that the vehicle was again stacked on the pad. There were an additional 15 days of tests and checkouts, and with a few cancellations because of weather and minor hardware problems, the prototype made its second flight after a total turnaround time of 45 days.

After the second flight, the vehicle was ready and on the pad in 20 days, but it sat there for another 30 days while the propulsion engineers waited for the results from the inspection of the first set of engines. These came back favorable, and the third flight was launched within 48 hours, using the same engines as the second flight. Based on the inspection of the first set of engines, Pratt & Whitney was willing to certify the RL60s for four flights with only an on-vehicle inspection between flights.

For the fourth flight, the vehicle was turned around and launched within 30 days, and by the fifth it was down to 20 days. Even though the engines on the first stage were showing relatively little performance loss between flights, it was decided to change them out while the second set of orbital-stage engines was pulled and sent back to the supplier for inspection. The first set of RL60s was back from Pratt & Whitney and was reinstalled in the orbital stage. Even with an engine change on both stages, the sixth flight of the vehicle was accomplished 30 days after the fifth. By working three shifts, the engineers and technicians responsible for each subsystem were able to get the next four flights off in a period of 35 days, for an average of just under 12 days turnaround. That was in spite of the need to replace one of the landing gear, which was pranged during landing on flight number eight. The fastest turnaround achieved, between flights seven and eight, was just under eight days.

In less than six months time, with nine successful flights, and an actual vehicle processing time of less than eight days between two of the flights, the major goals of the flight-test program—demonstrating true reusability and rapid turnaround—had been achieved to Forsyth's and Tom's satisfaction. At that point, further flight tests were put on hold while the orbital stage was thoroughly inspected, the engines were again returned to Pratt & Whitney for another teardown, and a third set of ablatively cooled first-stage engines was delivered and installed.

Following this three-month hiatus, the schedule called for a launch every two months. This would allow the flight operations team, now taking over from the engineers, to build their experience base and develop methods and procedures for eventual routine service. This phase of the flight-test program would also provide an opportunity to test various design modifications. To provide a backup, as well as to support possible additional test flights, the second prototype orbital stage was completed. Tom aimed to keep the flight operations budget below $100 million per year, so that he could concentrate the bulk of the available resources on finalization of the production vehicle design and the startup of production. Even with the test program delays, the first nine flights had been accomplished within three years from the provision of major funding, or almost four-and-a-half years since Forsyth had started to pull together the engineering team.

THE flight-test program was successful in more than a technical sense. Although it didn't convince everyone—or even perhaps anyone—that low-cost space transportation was finally just over the horizon, it had nevertheless demonstrated manned orbital spaceflight for an amazingly small amount of money. Sure, the prototype carried less than 1000 pounds of payload and the operations costs for each flight were over $10 million, but, compared to any previous manned orbital program, that was incredible. Based on the data from these flights, statistical analysis indicated a vehicle reliability in excess of 98%, with the distinct possibility that it would in actuality be quite a bit higher; Tom was aiming for three nines or better. Expendable vehicles, of course, often achieved 98% reliability, and the shuttle was inching its way towards 99% reliability, and so a reliability of 0.999 did not seem out of reach for the much simpler DH-1. However, it was the nature of these things that it would require hundreds of flights before a real idea of the DH-1's reliability could be known. But the relative simplicity of the system, and the low buy-in cost, would at the very least bring manned spaceflight to the point where almost any nation—and more than a few private companies or even individuals—could afford to join the club.

Following the successful nine-flight-test phase, and the commitment of the next round of funding, Forsyth was able to land firm orders for 12 vehicles, and he expected to be able to bring in another 12 before production began in 18 months time. The first and the eighth vehicles were reserved for internal developmental flight testing, meaning that AM&M would have a nearly three-year production backlog. Less than one-third of these orders came from the Air Force or NASA. Two actually came from private corporations and one from a private individual. The remainder came from foreign aerospace agencies in Japan, Germany, Great Britain, Italy, and Spain.

During the past year, the test pilots and crew systems people had been working to organize a spaceflight training school for customer pilots and crew. With the conclusion of the flight-test program, this school was now opened for business, and over 40 students were enrolled by the various customer organizations. In conjuction with the spaceflight training school, the operations and support group opened a sister school for ground crew, maintenance, and support personnel, with an initial enrollment of over 100 trainees.

Customers were not the only ones taking a serious interest in AM&M and the DH-1. It finally looked like Wall Street was ready to invest in a new space industry. Again, Forsyth—always wanting to get the money in hand before the inevitable problems arose—was willing to sacrifice equity for the additional assurance that large funding reserves gave to keep things rolling. He immediately began working with the underwriters to take AM&M public. During the three-month run-up to the IPO, no flights were conducted because of the possibility of an accident. However, investor Number 3, in cooperation with the underwriters, placed his stock for sale so that he could finance the lunar monastery. Contingent on the sale of his shares, he placed an order for two lunar transfer vehicles, at a total cost of just under $240 million. He also signed a joint endeavor agreement with AM&M and another company owned by Pete Van Horn, a space enthusiast billionaire whom Forsyth had not been able to recruit into the Group of Seven, but who was now interested in building an orbital refueling station.

It finally looked like Wall Street was ready to invest in a new space industry.

The underwriters made it clear that Number 3 was bailing out of his ownership for the sole purpose of purchasing vehicles from AM&M and to finance the engineering for the lunar transports and monastery, and the public offering went forward without a hitch. A total of $5 billion was raised, making up a 30% stake in the company. That meant $4 billion for AM&M and $1 billion for Number 3. On paper, this represented an almost five-fold increase in the value of their stock, which made the Group of Seven quite pleased, but not so pleased that Forsyth could convince them to sell even more shares to raise additional cash against unforeseen contingencies.

Forsyth had for some time now been carefully building an in-house sales team, which now comprised five top-notch salespeople. Most of their experience had been in the large aircraft market, military as well as commercial. They were gradually taking over his role of prospecting and working to close deals with various national, international, and commercial entities around the world who were interested or could be gotten interested in purchasing a vehicle. Over the next 18 months, as the DH-1 moved steadily towards production, a trickle at first and then a small flood of capital began to flow into other space-based business ventures that were interested in purchasing and operating a vehicle, or even several vehicles. Our friend Alexander Krempon was beginning to raise serious money to fund the planning and hardware development phase for his transpolar lunar expedition. But all of these would take time to come to fruition, and more than one would fold up before they ever placed a purchase order.

Over at the FAA, AST had reached the point where they were willing to grant blocks of launch licenses for up to five launches at a time. Processing launch licenses for the nine test flights in a six-month period had been exhausting on all sides. They were still far from willing to license the vehicle as a commercial, reusable launch vehicle, and they continued to study and draft proposed standards. At this juncture, Forsyth was content to be heavily involved in that drafting process and was not particularly interested in rushing things. It was clear that AST would become more willing and able to grant blocks of launch licenses for experimental and test purposes, and for the time being that was quite sufficient. With time, more and larger interests groups would bring pressure to bear as a growing and increasingly diverse list of operators acquired DH-1s and wanted to use them on a regular basis. At the same time, a longer track record and a greater comfort level with the whole process of turning the vehicles around between flights would make it easier for AST to approve their routine operations.

Of course, although passenger flights were still in the future, Forsyth was not hesitant to offer the copilot seat to pilots from organizations that were interested in purchasing the vehicle and even, on occasion, to individual customers if they wanted a test drive. This required signing them on as "mission specialists," and he certainly couldn't charge them for the flights. Nevertheless, in the month following the resumption of test flights, they flew an Air Force colonel and an *Aviation Week* editor. AST did not object.

CHAPTER 26
Money, Manufacturing, and Marketing

TO not waste any time looking for a new site, and to keep it near the flight-test operations center, the production facility was located adjacent to the existing prototype fabrication facility at AM&M's Florida campus. Additional funding from the state of Florida paid for more than half of the cost of the facilities expansion. The new plant was designed for a production rate of up to 12 vehicles per year. With a startup rate of eight per year, however, and an expected build time of six months, there would initially be only four vehicles in process at any given time. The actual production space was less than 100,000 square feet, with an additional 40,000 square feet of environmentally controlled storage space for the orbital-stage propellant tanks. These would be shipped in groups of six from Japan. Even though only four vehicles would be in process at a given time, that meant there would be eight stages under construction. So far, all of the vehicle orders—except for the lunar transport ships—were for both stages. As the market developed, Forsyth expected that the number of first stages built would fall to less than 50% of the number of orbital stages.

With the close of the second round of financing, Charlie Luben, AM&M's facilities manager, had the contractors pouring concrete night and day, and by the end of the ninth test flight the shell of the production facility was complete. During the relatively quiet period during the "due diligence" phase of the initial public offering, Tom himself gave me an tour of the production facility. I was surprised to find that the production lines for the first stage and for the orbital stage were completely separate. In fact, it was not possible to get from one production line to the other except through the engineering offices. There were a few large doors in the wall separating the two lines, but these were all locked. I asked Tom why the lines were isolated like that.

"I used to buy components from a company up in New Hampshire that made gyroscopes and steering actuators in the same facility, side by side. The gyros were high precision and high-tech, while the actuators were by comparison rather low-tech. Over time, they found that either the quality of the gyroscopes suffered, or the cost of the actuators started to ratchet up, as the same manager and supervisors tried to run both at the same time." He paused, then went on when he saw my quizzical look.

"There's a tendency for a single standard of production to prevail in a single facility. That standard may not be reasonable for all products, depending on their complexity or required quality level," he explained.

"But aren't the first stage and the orbital stage on the same level of sophistication?" I asked. "They're both manned rocket vehicles, with the same safety and reliability requirements."

"Yes," agreed Tom, "that is certainly true. However, the criticality of weight in the two stages is different by a factor of almost 20 to 1."

"I think I see what you're getting at," I said after a moment.

"Remember, for the orbital stage, removing a pound of weight, even if it increases the cost of the production vehicle by $10,000, is well justified," Tom continued. "We're still working our way toward the 5000-pound payload we want, and a pound of vehicle weight eliminated means another pound of payload. So that has to be constantly on the minds of everyone involved in building the orbital stage, no matter what part they play. Whenever bonding agents, fasteners, connectors, wire and cables, coatings, and finishes are chosen and integrated into the vehicle, it needs to be done with the perspective that it's usually OK to trade cost for reduced weight. In the first stage, on the other hand, we can generally solve problems by adding weight—if that proves cost effective. The first stage basically involves a technology level and weight sensitivity that is on par with transport aircraft. By separating the production lines, the workers, and at least the first tier of supervisors, I hope to avoid unnecessary weight in the orbital stage, and unnecessary cost in the first stage." Tom made it clear that he intended to keep those doors locked.

After the tour, Tom took me to meet Oscar Martinez, the chief manufacturing engineer who, although he was an engineer and had an engineering title, was heavily involved in coordinating the activities of the many subcontractors and the major partners, which included Pratt & Whitney for the first-and second-stage engines, Garret Air Research for the fluidics system and control actuators, Pioneer Aerospace for the parawing and drogue systems, and, most particularly, Mitsubishi Heavy Industries for the orbital-stage tanks and structure. He also found himself working closely with Bechtel on the launch facilities and ground handling equipment

Oscar gave me a thorough overview of the manufacturing challenges that the DH-1 presented. The orbital stage was critically dependent on control of the processes used in its construction, especially with regard to the propellant tanks. That was one reason Mitsubishi had been the hands-down winner of that subcontract: Japanese companies are renowned for their ability to achieve, with great persistence and care, consistent results from difficult industrial processes. For the tanks, there were four major issues. First, the aluminum alloys were specified with unusual levels of purity for the aluminum and the alloying elements. Second, the tank wall thicknesses were specified to plus or minus one-thousandth of an inch throughout the entire vehicle. Third, the overall dimensions of the vehicle had to be controlled to within about 50-thousandths of an inch. This ensured that mass properties and aerodynamic properties would be consistent between vehicles. But more importantly, the entire vehicle structure was to be loaded to 100% of the elastic limit (while filled with cryogenic fluids) so that any stress concentrations would result in local yielding. That would put existing cracks that were too small to detect or any other stress concentrations into compression. This necessarily caused the vehicle to stretch a bit, and the tight tolerances on the overall dimensions were to ensure that this occurred uniformly between vehicles. None of the airframes for the first three prototype vehicles were exactly the same, because of this stretching, and therefore their overall dimensions and mass properties were different. Tom wanted the production tanks built with this unusual level of control so that the preloading operation would produce the same deformation in every airframe.

A fourth critical operation was the installation of the cryogenic insulation. This consisted of blocks of open-cell polyimide foam bonded to the inside wall the hydrogen tank. The bonding agent selected was generally highly reliable for cryogenic applications, and in laboratory tests with small samples over 1000 cycles had been achieved without delamination. However, when scaled up to a full-size tank, the original process left something to be desired. Within a half-dozen cycles, large areas of the insulation had to be ripped out and replaced. In addition, with over 98% of the welds performed using stir-welding, which in and of itself is a relatively new and exacting process, great attention to detail was needed in this area. After the 100% elastic, or strain, loading was completed, the vehicle was inspected using thermal flow techniques to detect cracks down to 1/50 of an inch. The tank was then placed into a positive-pressure container for shipment to the assembly plant at the Cape.

A complete, finished tankset weighs about 6000 pounds, and Mitsubishi had invested over $300 million in the construction of a new factory and a low-contaminate aluminum casting and rolling facility. So, even though aluminum normally costs $2 to $3 a pound, and the special marine alloy used in the DH-1 not much more, the completed tanks ended up costing just over $1000 a pound. But Tom and Oscar considered them cheap at that price. The first 100 tanks were to be built using the high-purity marine alloy. Beyond tank number 100, Tom planned to go to an aluminum-lithium alloy having the same or better structural and thermal properties, but at a density about 10% less than that of the marine alloy. At this time, the aluminum-lithium alloy was only available in pilot plant quantities and wouldn't be ready for production—to Oscar's satisfaction—for another few years. The lighter alloy was expected to yield a roughly 600-pound gain in payload. AM&M had already signed a letter of intent to buy the aluminum-lithium tanks at a cost of $12 million each.

For the other major components of the orbital stage, costs were as follows: the twin RL60 engine system cost about $12 million per shipset; the cold-gas reaction control system (RCS) and hydrogen peroxide orbital maneuvering systems, just under $1 million; and the avionics suite, including the flight computers but not the software, came to $150,000. The parawing and related systems were currently at $250,000, but with a planned switch in the next year or two to a new, high-strength fabric—made either from Spectra or Cuben fibers—that was expected to increase by a factor of five. For that, they would get a weight reduction of 50%, and the useful life was expected to grow from 5 to over 20 flights. Each orbital stage came with one ejection seat as standard equipment, although almost all orders were specifying two or three seats, the maximum currently available until additional flight experience was gained. Each ejection seat cost about $250,000 installed. The heat shield was coming in at less than $500,000, but it was not yet clear how many flights it would be good for. So at this point in time, the marginal cost of building an orbital stage was less than $50 million, and in fact, if you looked at the suppliers' marginal costs, it was quite a bit lower still. The first stage, in the all-rocket configuration, had nine engines at an average cost of $1 million each. The rest of the first stage, with its relatively low-tech aluminum tanks and simple structure, cost less than $5 million, including the avionics and ejection seat. Thus the marginal cost of a complete DH-1 was only about $65 million.

AT a production rate of eight vehicles per year, a list price of $250 million per vehicle would bring AM&M a tidy marginal profit, before taking into account the development and other fixed investment costs. Forsyth, though, was intent on reinvesting most of the profit in quality and performance improvements for the vehicle. The goal remained 5000 pounds of payload, at least once the improved material for the parawing was flight qualified. That figure was based on a single ejection seat and only about 200 feet per second worth of orbital maneuvering propellants. With the orbital maneuvering systems tanks full, the original nylon parawings, and a second ejection seat, the payload was closer to 3000 pounds. Still, none of the customers to date were particularly concerned about the payload capacity. Rather, they were satisfied to be acquiring a reusable, manned space launch system, at what appeared to be a bargain price. Even if you didn't believe that the vehicle had long-term viability, it seemed a cheap way to gain experience with manned space operations and reusable launch vehicles.

Nonetheless, Forsyth was interested in pursuing enhancements for the DH-1 in the near term. The construction of the first stage, which operated within the atmosphere with all of the possible hazards that entailed—from lightning strikes to wind shear—was particularly robust. That, along with the weight insensitivity of the stage, provided opportunities for modifications that would allow boosting larger and heavier orbital stages for certain missions. One such modification involved the addition of strap-on propellant tanks. This would reduce the thrust to weight of the vehicle to as low as 1.15, but would allow, with the use of a downrange trajectory rather than the typical vertical boost trajectory, launching an orbital stage with 10,000 pounds of payload or a modified stage with a total dry weight of 22,000 lb.

One of the potential applications for the strap-on version would be the launching of a space station or refueling depot in sections. This would be accomplished by mounting the modules to the forward portion of the orbital stage. The trajectory would be designed to keep the vehicle stack subsonic below 80,000 feet, thus greatly reducing aerodynamic drag and maximum dynamic pressure. This would allow the relatively lightweight modules to be mounted to a simple support flange, tied into the forward payload bay attach ring.

One of the potential applications for the strap-on version would be the launching of a space station or refueling depot in sections.

Furthermore, although the initial run of 16 first stages was to be built with the hydrogen peroxide landing rockets, the basic airframe was designed to accommodate four or six jet engines in place of the rockets. These engines were the new design being developed by Pratt & Whitney at AM&M's behest, based on a Russian-designed liftjet, the Kolesov RD36V-35FV. It is a conceptually simple engine with an excellent thrust-to-weight ratio of almost 15. It was looking like the liftjets for the landing system would come in at a weight of 3000 pounds, with a maximum thrust-to weight of about the 15, which had been the original goal. However, the baseline 30,000 pound thrust would be boosted for touchdown to around 45,000 pounds with water injection. The complete jet engine installation, including mounts and intake doors, plus 5000 pounds of fuel would weigh a total of 10,000 pounds, which was very close to the weight of the rocket system and its propellants. The big plus, though, was that with a fuel consumption of 4 pounds per second this system would allow some 1000 seconds of maneuver and hover time at nominal thrust, and 50 seconds with water injection. That was a vast improvement over the 10 seconds available from the rockets.

While the engineers were hard at work on these projects, the outside world was only slowly beginning to respond. Although the popular press and the space enthusiast crowd were waxing eloquent about the promise of the DH-1, there was as yet no indication of any collapse of the expendable launch vehicle market, or the grounding of the space shuttle before its (ever-receding) official retirement date. First production vehicle delivery was still over a year down the road, and the industry pundits were still skeptical of the small payload—at least in terms of typical military or commercial space systems, particularly geosynchronous satellites—and the unremarkable operating costs that had been demonstrated by the AM&M vehicle to date. And even Forsyth would admit that expendable launch vehicles might have 5 or even 10 years of life left. No matter how fast or how far launch costs fell, it would take time to build an infrastructure that could support the launch of major communications or military satellites in pieces, with assembly and refueling on orbit, followed by boost to geosynchronous orbit. And if it looked like 5 or 10 years to Forsyth, undoubtedly it looked more like 15 or 20 years to the industry establishment. Thus, there was no concerted effort to hinder the development or to prevent the export of DH-1s, except on national security grounds. AM&M continued to work closely with the National Security Agency, the State Department, the Commerce Department, and FAA/AST to jump

While the engineers were hard at work on these projects, the outside world was only slowly beginning to respond.

through all of the hoops, build all of the coalitions, and offer all of the assurances necessary to open up the market to every nation that was not at war or on the list of terrorist-supporting states.

That was exactly as John Forsyth had hoped it would be at this point—the sleeping giants of the aerospace community looked like they would stay asleep for a few years yet. NASA, although they had committed to purchasing a vehicle, had no current plans to make it compatible with the International Space Station (ISS). There was, however, some movement within the agency to consider keeping it ready to launch as a rescue vehicle for the space shuttle or the space station. But these plans were not far along.

The arrival of the AM&M vehicle diverted some attention away from NASA's ongoing "Third Zone" launch vehicle effort, which was looking more and more like another short-lived development program that wouldn't produce an operational system. NASA was beginning to feel a budget squeeze even though funding for this system, with its Mach 6 first-stage launch platform, had not yet reached its peak. Signs of programmatic trouble had been brewing for some time. Nonetheless, the aerospace establishment was still busily at work on its new technologies and state-of-the-art hypersonic systems. To NASA and the big companies, the DH-1 seemed dull and uninteresting by comparison, and certainly not a threat.

No doubt there were some, particularly in Congress and the military, who could see glimmerings of what Forsyth had been working so hard to bring about—a paradigm-shattering change in the way things were done in space. But for most, there was really no desire for the approaching revolution, even if they could see that it might be coming. Their programs were funded, political consensus had been built, money was flowing to companies, research centers and universities throughout the country, and the last thing they wanted to do was provoke a debate on the future direction of space launch operations if there was any possibility of a radical change.

The very low end of the launch market had seen the bankruptcy or shutdown of several companies, which had not, in any case, been profitable, or in several cases had not even achieved a successful first launch. Although it was not clear that the approaching availability of the DH-1 had been the primary cause of their demise, it was probably safe to say that the approaching availability of routine space access, at least for small payloads to low Earth orbit, had undermined the reason for existence of several of these small ventures. They perhaps ran out of funding or called it quits a year or two earlier than they otherwise would have. And at least one small launch-vehicle company had placed an order for a DH-1, apparently hoping to service their existing customer base with it.

CHAPTER 27
Always Room for Improvement

IT had been some time since I had talked with Paul Reston, the performance analyst. During my latest talk with Tom about the performance improvements that were in the works, I had expressed my concern that the prototype vehicle came dangerously close to having no payload at all, and to my mind the current first-run production vehicle target of 3000 pounds seemed far less than necessary to get space transportation costs to start falling precipitously. Tom suggested that I go talk to Paul about the improvement program; one of his jobs was to look at the vehicle and to determine where increases in performance or decreases in weight would provide the greatest payoff.

Of course, it was obvious that a pound eliminated in the orbital stage was a pound of payload gained. As Paul explained, however, there are some other subtleties having to do with engine performance, residual propellant management, tradeoffs between a larger parawing and a more sophisticated configuration with better lift, software to control the landing, and many, many more. Paul had been tasked to assign a dollar value to this host of possible incremental improvements so that Tom and his engineers could determine which ones to pursue or to ignore, based on a cost-benefit ratio. To do this, Paul had to determine, or at least estimate, what a pound to low Earth orbit was actually worth. The real price of space transportation had been coming down slowly in the last decade, and by some measures was now below $3000 per pound to orbit. Forsyth, of course, had pegged a cost of $200 per pound as the near-term goal for the first-generation AM&M space transport. So the first number in Paul's calculation could vary by more than a factor of 10.

The number of flights for each vehicle over its lifetime and per year were the next numbers that Paul needed, and these too embodied a good deal of uncertainty. The number of flights each vehicle would be able to make during its lifetime of course affected the amortization of the cost of the vehicle, and also determined the number of flights that could be expected to benefit from any given performance improvement or weight savings. Currently, AM&M was guaranteeing vehicles for 10 successful flights, and the expected flight rates were two to five flights per vehicle per year in the short term. The minimum design lifetime goal for the DH-1 was over 1000 flights for the basic airframe. The engines were currently only certified for 10 flights before tear-down and overhaul, but it appeared that they were inherently capable of 25 flights between overhauls, with an operational lifetime of more than 100 flights for a ship set. There didn't seem to be any showstoppers for running the engines for hundreds, perhaps even 1000 flights. At this point, though, they simply didn't have the operational experience to make an accurate assessment of engine life. There was even less to go on for the heat-shield system. Laboratory tests and simulations were showing encouraging trends, but they would have to have real-life data before they could be sure. For the avionics systems, it was presumed that even though they might very well last for decades, just as in the case of commercial aircraft, they would nevertheless be replaced in large measure every 10 to 15 years. The continual evolution of new generations of computers, instruments, and displays meant that it would not be cost effective to maintain the existing systems past a certain point.

In the near term, and considering the cost of money, the total number of flights a vehicle might eventually achieve was less important than the number of flights it made per year. With all of this in the equation, Paul took as a reasonable estimate for the total number of flights over which the return on investment in a performance-enhancing project could be realized as five times the number of flights each vehicle could be expected to make in a single year. The DH-1 was certainly, from an operational point of view, capable of being launched more than once a day. It seemed though, that for the foreseeable future most of them were unlikely to be launched more than half a dozen times a year. So, the second number in Paul's equation—the number of flights per year—could vary by nearly a factor of 100. But to a certain extent, the cost per pound to orbit and the number of flights expected per year balanced each other out. If annual flight rates per vehicle were down around two or three, a pound of payload to orbit was likely to have a value in the $1000 to $2000 range. On the other

In the near term, the total number of flights a vehicle might eventually achieve was less important than the number of flights it made per year.

hand, if the annual rate went up to 100 flights per vehicle per year, it would probably mean that something like Forsyth's target of $200 a pound had been achieved.

The magnitude of the space lift capability that the DH-1 promised was truly awesome. If AM&M produced only eight vehicles per year for the next five years, and if by that time each vehicle was flying only 12 flights per year, that would amount to 480 flights per year times 5000 pounds of payload per flight, or more than two million pounds per year to orbit. That was more than four times the current annual mass to orbit, if you were very generous with what you considered payload. So, if AM&M could manage to stay alive at even that low production rate for the next five years, it would enable a factor of four growth in launch traffic. Whether or not the market could expand to absorb that capacity would depend on many factors beyond AM&M's control. Actually, Forsyth was hoping for a lot more growth than that.

AS always in engineering, and particularly when costs and economics are considered, a large measure of judgment is needed. Paul's approach was to come up with bounds for the recurring costs and for the nonrecurring, or development costs, that were conservative but at the same time would provide useful guidelines for planning a continuous improvement process for the DH-1. So whether it was 10 flights at $1000 a pound or 100 flights at $100 a pound, it seemed that a reasonable estimate of the dollar value of increasing the payload of the vehicle by 1 pound was on the order of $10,000. Next, he estimated the value of making a design or production change that would enable all future DH-1s to carry an extra pound each into orbit at 10 times the cost of improving a single vehicle. Thus, AM&M should be willing to spend an additional $10,000 in the purchase price of a component, if the additional cost would result in one more pound of payload capacity, and willing to invest $100,000 in nonrecurring costs if it would improve the vehicle design so as to yield an extra pound of payload capacity on each subsequent vehicle. For example, if an actuator could be replaced with another one that weighed a pound less, AM&M should be willing to pay $10,000 more for the lighter one. Or, if a structural component could be redesigned to reduce its weight by 1 pound, the company should be willing to invest $100,000 in engineering and other development costs. Of, course, there would typically be some nonrecurring costs involved in switching components, and a redesign was likely to result in some increase in recurring costs. But the conservative numbers that Paul had settled on should provide sufficient margins for covering both the nonrecurring and the recurring costs for a weight reduction or performance improvement effort.

Of course, reducing the weight of some subsystems could be particularly advantageous. For instance, there were two engines, so that eliminating a single pound from an engine would generally produce 2 pounds of additional payload. Thus, if it were possible to reduce the weight of an engine from 800 pounds down to 700 pounds, it would justify a $20 million investment, and a $1 million bump in the price of each engine. For the cold-gas reaction control system, which included 12 thrusters, reducing the weight of each thruster by a pound could justify a more than $1 million per pound investment in engineering and development and a $10,000 price increase per thruster, for a total of $120,000 in additional procurement costs.

The payload capability could also be increased by improving the vehicle's performance. As Paul explained, a 1-second improvement in the specific impulse of the engines would provide a 70-pound increase in payload performance. Or a 1-foot per second decrease in the velocity needed to achieve orbit—say, by reducing steering losses—would result in a 1-pound gain in payload, and vice versa. Numbers like this are also convenient for doing quick mental calculations to estimate the amount of payload that could be lifted to a higher orbit.

Some of these improvements, such as producing a lighter-weight or higher-performance engine, had lead times of three or four years. Others could be developed and implemented much more quickly. But when I talked to Oscar Martinez about incorporating these improvements in vehicles on the production line, he made a point of emphasizing that when you are building safety-critical products, configuration control is very important. In other words, everyone involved in the process has to know exactly what is supposed to be built at all times. Known critical processes, procedures, and specifications can be dealt with through training and procedures, as well as with familiarity. But every design or component change will introduce new requirements—and new potential failure modes—that need to be understood, documented, disseminated, and trained to. To keep the DH-1 production process firmly in control, individual components would be upgraded only at set milestones. Consistency and therefore safety would be aided by this conservative approach to modifying or upgrading the production vehicles; once a particular change was approved and implemented, it would become the standard for all future production vehicles. Any major changes to the structure or key subsystems that were not going to be incorporated in all of the production vehicles would not be made on the production line.

Oscar explained that the special stretched versions, such as the lunar transports, would be completed by taking a nearly finished production vehicle and remanufacturing it on a separate line, where a new payload compartment, the lunar landing gear, and other special equipment would be fitted. In this way, the possibility of confusion and mistakes on the production floor was minimized. Oscar told me that this had been the practice for aircraft production during World War II. Once an airplane factory had started production, the design produced in that plant was not changed. The airplanes were simply shipped to a second factory where they were remanufactured with the latest upgrades. In this way, the productivity and experience of the original factory and its workers were not compromised, and production did not have to be stopped while the upgrades were integrated into the lines. This approach had the additional advantage of forcing the engineering team and the suppliers to think of design improvements in terms of how easily they could be retrofitted into existing planes.

Of course, as Forsyth had expected, the Air Force and NASA purchase orders were already specifying modifications to the basic DH-1. The problem with government procurement is that any modifications to the vehicle rendered it no longer a catalog item, and so the government by law had to pay the actual cost and not the market price. Forsyth was not about to give the government the detailed cost accounting data that they would need to confirm the "actual cost"—he regarded that information as a very valuable company trade secret. Other difficulties would arise if AM&M were selling basically the same vehicle at significantly lower prices (or even higher prices) to other customers.

The solution had been to facilitate the formation of a separate company, which entered into a joint venture with one of the major aerospace companies. This other company, Federal Space Systems, bought standard configuration DH-1s from AM&M at catalog prices, and then performed all of the modifications necessary to bring them up to Air Force or NASA specifications. The vehicle and the modifications were then billed to the government on a strictly cost basis, plus a "reasonable" profit, as defined by the government.

Indeed, Forsyth hoped to sell quite a few DH-1s to the government in this way. He had been working for some time with a number of enthusiastic Air Force officers who were out to define a new Air Force space defense doctrine. The cornerstone of his approach was to subtly push the idea that the Air Force ought to build a strong, dominating—but unarmed—presence in space, while developing space-based weapons that could be stockpiled and mounted on their space assets in the event of hostilities. In this, he had apparently been successful—or perhaps the Air Force had been headed in that direction anyway. Several senior officers were now actively moving the program forward, garnering congressional support and deflecting international objections, with a plan to achieve a major role in space while avoiding the charge of militarizing space. Their goal was to be seen as the friendly cop on the beat, ever present but not threatening to those who meant no harm. Forsyth was of course hoping for a new space race, but certainly not an arms race. No, he wanted nations to strive for presence, commerce, and the glory of exploration in space; and, yes, for participation in space traffic control and information gathering, but not for military dominance. He did not want the advent of the DH-1 to be seen as a step toward American militarization of space. He wanted it to be the means by which all people and all nations could participate in creating in space a prosperous and open, but also secure, infrastructure—and then, an ever-expanding frontier.

It remained to be seen what would follow. Would Forsyth's vision become reality, or fade away like those of so many other space dreamers? Only time would tell.

EPILOGUE I
Space Is Finally a Place

I again take up the chronology of AM&M after an interval of about 10 years. A number of things have distracted me from keeping up with the story during that time, including several short books and a screenplay that John Forsyth wanted to help promote the vehicle. By the time I completed those, the DH-1 had been in production for over a year, and it was already becoming apparent that Forsyth had been successful in changing many of the basic assumptions about space. Soon, people stopped wondering whether you could make money in space—the new question was "hey, how can I get in on this?" Five years after production had commenced, AM&M's order backlog had climbed to almost 100 vehicles, and over 60 vehicles had been delivered. Then a worldwide recession coincided with the loss of three DH-1s in rapid succession, and together that was sufficient to pop the space enthusiasm bubble.

Fortunately AM&M, like the better-managed high-tech companies of the 1990s, had squirreled away a large nest egg of cash. In the first five years, Tom and Oscar had been able to cut the cost of production by a third, and during the recession the sales price of the vehicles had been reduced by nearly half. Even so, unit sales after the pop fell from a high of 24 a year to less than 16, and the backlog essentially evaporated. Nevertheless, enough vehicles had been sold to make Forsyth's plan of creating a huge overhang of launch capability a reality, and in the face of the recession operators were struggling just to cover their marginal costs.

Before the downturn, a joint venture company was organized by AM&M, Number 3, and Pete Van Horn to build a refueling station in low Earth orbit. They raised over $2 billion and built an equatorial launch site in Ecuador at about 5000-ft altitude, where the climate was tolerable and the necessary industrial base could be supported. They had also purchased three tanker versions of the DH-1, designated the DH-2, and had contracted for the launch in sections of an on-orbit hangar and the refueling depot. It was designed for long-term storage of over half a million pounds of liquid oxygen, liquid hydrogen, hydrogen peroxide, and liquid methane. Funding for all of this had been relatively easy to come by during the heyday of the boom, but the demand for orbital refueling had not appeared fast enough, and the company went broke when the bubble popped.

Number 3 organized a rescue effort that bought the assets of the bankrupt company for pennies on the dollar and was now prepared to be the major customer. He and Father Scipio were finally ready to begin construction of the monastery at the lunar south pole. Alexander Krempon, who had been planning an expedition to the moon, had also raised a considerable amount of cash during the boom. He persuaded Number 3 that he, Krempon, ought to be the first to return to the moon, with an expedition that would lay the groundwork for both the monastery and his own transpolar expedition. However, in a bold ploy the likes of which was inconceivable in the era of government-only spaceflight, Krempon's ship had boosted not for the moon, but for Mars. No one had questioned the modifications that had been ordered for the ship—it was certainly plausible that he needed the extra propellant capacity for extended lunar operations or contingencies. He of course had neglected to modify his flight plan before departure, and the United Nations (UN) committees that were still debating the issues of contamination and protocols for a manned Mars expedition were outraged. However, the public excitement and support that the three-man mission to Mars engendered around the world ensured that by the time Krempon and company landed on Mars 256 days later, the UN had little choice but to retroactively grant permission to land, based on the new facts generated by the expedition.

After his triumphant return from Mars almost two years later, Krempon had no trouble raising the necessary funds for his lunar transpolar expedition. Krempon's voyages really brought out the drama in space exploration and also spurred many other such efforts. Even before he made it back to Earth, five western nations plus Russia and China had all determined that they needed a base on Mars. This led to a wave of major expeditions there, the first of which occurred about a year after the successful conclusion of the transpolar expedition on the moon.

With that, things finally started to heat up again. Orders for over 70 DH-1s were placed in the next two years, including three additional tankers for the orbital refueling operation. For the past two years, there had been six regularly scheduled supply flights to the moon each year, meaning that you either stay for a day or at least two months. The main cargo consists of supplies for the monastery, a small Air Force research station, and the international science base. With increasing frequency, other scientific expeditions that operate from the bases at lunar south pole have been organized and carried out. Perhaps the most amazing development is the ultimate adventure vacation: a flight to the moon with a stay at the monastery, for a fare of $2 million per person. The accommodations at the monastery are spartan but free for as long as you want; however, the monks tend to put visitors to work after a while, doing light gardening, washing the dishes, and so on. This is often exactly what the millionaires who make the trip find most restful, and although the monks don't charge for the stay, the guests are usually quite generous upon their return to Earth.

Perhaps the most amazing development is the ultimate adventure vacation: a flight to the moon with a stay at the monastery, for a fare of $2 million per person.

The most recent count shows that over 250 DH-1s have been built, with the vast majority still in operation. Average utilization for the fleet is now almost 25 flights per year. The major sources of demand are an array of low-Earth-orbit outposts operated by more than 12 nations and a growing number of geostationary space stations. These include scientific and research platforms, military command and control centers, and space traffic control centers, covering the entire inner solar system. The latter function has grown rather dramatically, given the surge in trips to Mars, and the surprisingly large volume of traffic to the more accessible asteroids. More than half a dozen companies or individuals have managed to raise venture capital for expeditions to explore and de facto claim various asteroidal bodies, which, it is hoped, will be worth trillions in mineral and water resources. A consortium of U.S. and Russian companies is producing nuclear power plants with outputs ranging from 250 kilowatts to 2.5 megawatts, which can be launched into space and transported to the moon, Mars, or the asteroids to supply basic power needs.

With the growth in interplanetary traffic, flight rates for the refueling tankers are now over 250 flights per vehicle per year. The cost of a pound to orbit is less than $100 for bulk cargoes such as propellants for the refueling depot. All told, annual traffic to orbit has grown to more than 6000 flights, carrying over 30 million pounds of payload. The loss rate for the vehicles is on the order of once every 3000 flights, a notable improvement from the 1 per 1000 flights experienced during the first five years of operations. An international space transportation safety board, chaired by FAA/AST with participation from the aerospace establishment, including the air forces of the major world powers, carefully investigates each loss, and the causes of the failures are being eliminated one by one.

SPACE tourism, which for years had been touted as the "killer app" for low-cost space transportation, still amounts to less than 10% of overall launch demand, although that is growing at 10 to 30% a year, depending on the state of the economy. There is still widespread belief that it will become a major driving factor. A trip to orbit from the United States still requires an ejection seat, which historically doubles your chances of surviving a launch disaster. The ejection seat requirement has kept the cost of a ticket to orbit at about $200,000 because the DH-1 can only fly five passengers at a time that way. At those fares, although most of the vehicle operators are covering their marginal costs, a trip to orbit is still being sold at less than actual cost—if the cost of capital is included. In Asia, where as many as 20 passengers can be stuffed onboard, a flight to orbit can be had for $50,000. A week's stay on orbit will set you back an additional $50,000 in one of Bigelow Aerospace's two still-Spartan low-Earth-orbit hotels. There are numerous plans in the works for more luxurious accommodations, but the financing has not yet materialized.

The current annual market for transportation to orbit is approximately $6 billion. According to some estimates, another $100 billion is being spent on the payloads that are being launched into space, and the entire industry including related manufacturing and ground support is in the neighborhood of $500 billion. AM&M, which is now producing 30 to 40 orbital stages and 15 to 20 first stages a year, has annual revenues from vehicle sales of between $4 and 5 billion. Additional revenue from spare parts comes to another $1.5 billion, exclusive of ground facilities. Because the cost of the vehicles and spares equals or exceeds the total revenues from flight operations, it seems likely that more than a few vehicle operators are not making any money. But this is somewhat misleading. Many DH-1 operators, including the armed services and various governmental entities, are not trying to operate at a profit. Further, the vehicles are a very expensive capital item, and the total payback might be five or seven years into the future.

The cost of the vehicles has also continued to drop. Currently, an orbital stage, in the standard configuration, is selling for less than $100 million, and a first stage goes for less than $60 million. One first stage can support at least two or three orbital stages. The vehicle performance has also been improved substantially, mainly through improvements in engine performance, but also partly because of reduced empty weight. The latest model, the DH-1-700, can lift just under 7000 pounds to the altitude of the major refueling stations at 150 nautical miles, and almost 6000 pounds to the typical space station altitude of 400 nautical miles.

AM&M is two years away from the production of a larger vehicle, the DH-3. It will be able to place 20,000 pounds into low Earth orbit, using five RL60 engines in the orbital stage and six upgraded boost engines in the first stage. This orbital stage is being designed to accept three new, higher thrust engines, which Pratt & Whitney has in the early stages of testing, funded partly by the Air Force and partly by NASA. Even so, there are no current plans to stop production of the DH-1. Forsyth, well aware that many aerospace vehicles have had useful lifetimes in excess of 50 years, figures that as long as they keep making them and gradually lowering the price, it will be a long time before anyone else puts up enough cash to get into the business. He does not doubt, though, that eventually Russia, China, and maybe India will do it simply for the sake of national prestige, much like Airbus had been created by the Europeans to get into the commercial aircraft business. Still, serious competition for AM&M seems to be at least 10 and perhaps 15 years in the future. AM&M has a huge head start, covering the spectrum from basic orbital transportation to tankers and specialized deep space ships.

THE rapid growth in the worldwide launch rate, now in excess of 6000 launches per year, finally raised some environmental concerns both for the Earth's atmosphere and the low-Earth-orbit environment. By way of comparison, the launch of a DH-1 involves burning about 1/20 the amount of propellant burned in a single shuttle launch. The first stage burns relatively low particulate-generating methane and oxygen, in contrast to the very dirty and chemically hazardous ammonium perchlorate, hydroxyl terminated polybutadiene, and aluminum powder of the shuttle SRBs. Thus each DH-1 launch generates about 1% of the pollution resulting from a shuttle launch, but the annual impact of 6000 launches is about equal to that of 60 shuttle launches. As space transportation costs continue to fall, AM&M's economists are predicting an increase of at least another factor of 10 in launch rates. Certainly, if the growth of space tourism continues at current rates and major industries in space arise as asteroidal resources become available, 300 million pounds to orbit each year is not out of the question. Of course, in the next five years it is expected that enough water from the asteroids will be available in low Earth orbit so that it will no longer be necessary to launch propellants from Earth. In the long run, it seems clear that transportation of people to orbit will dominate the market.

The primary pollutants produced by the orbital stage are the free hydrogen in the exhaust and, somewhat surprisingly, the water vapor combustion products themselves. The flight trajectory is such that almost all of the hydrogen and water vapor of the orbital stage are released above 40 miles altitude. Because almost all of the propellant in the orbital stage is expended above the bulk of the atmosphere, a relatively small amount of hydrogen falls back into the lower atmosphere, where hydrogen naturally accounts for about one part per million. That works out to about 400,000,000,000 pounds of hydrogen. A single DH-1 launch with engines using a mixture ratio of 6:1 results in the release of about 2400 pounds of free hydrogen. Of these 2400 pounds, only about 40% are produced in the lower atmosphere. Thus, assuming 60,000 launches per year, the yearly total is 1000 pounds times 60,000 launches, which equates to about 0.006 parts per million excess hydrogen. That appears to be relatively insignificant.

Of greater concern is release of hydrogen in the upper atmosphere. The stratosphere and mesosphere contain about 66,000,000,000 pounds of hydrogen; 60,000 flights per year would release 1 or 2% of this amount. However, hydrogen molecules are constantly being stripped from the top of the atmosphere as they are heated and eventually achieve velocity sufficient to escape Earth's gravitational clutches. The total loss of hydrogen from the upper atmosphere above 84 miles is some 200,000,000 pounds per year. A flight rate of 60,000 per year would actually dump about half of this amount into the upper stratosphere, which is enough that it just might be significant. Forsyth has funded research at several universities to look thoroughly into the effects of such an increase in hydrogen in those regions.

The water-vapor issue has already received attention because it is also produced by high-flying jet aircraft. Commercial aircraft generate about 80,000,000,000 pounds of water in the lower stratosphere at an average altitude of 35,000 feet. The projected 60,000 flights of the DH-1 would release about 5,000,000,000 pounds of water during first-stage flight, primarily below 100,000 feet. Thus, the water vapor generated by DH-1 flights in the critical lower stratosphere, where cloud formation is stimulated by the contrails, would probably be less than 1% of that from commercial aircraft.

Still, if it were deemed necessary, the amount of hydrogen released could be reduced by increasing the mixture ratio of the orbital stage engines so that they burn oxygen rich. Actually, the second-generation engines, which could be retrofitted to existing vehicles, are being designed to operate at mixture ratios of up to 9:1. Although this decreases the specific impulse of the engines, it also increases engine thrust-to-weight because of the higher average density of the propellants. Overall, with the higher propellant load and higher thrust the performance of the orbital stage will come out slightly better than with the current RL60 mixture ratio of 6:1. It will also increase the gross weight of the orbital stage somewhat, but that can be compensated for by increasing rather marginally the propellant load and thrust of the first stage. For deep-space operations, the interplanetary vehicles will continue to use the 6:1 ratio in order to minimize the total propellant weight.

The particulate matter introduced into the upper atmosphere by the DH-1 is minimal. The first stage consumes most of its propellant below 60,000 feet and produces less particulate pollution than a jet aircraft of comparable thrust operating at the same altitude. The first stage consumes 168,000 pounds of propellants, of which only about 35,000 pounds is methane, with roughly half the carbon content of jet fuel.

If at some point in the future it appears that space launch operations are introducing too much particulate matter into the upper atmosphere, hydrocarbon fuel could be eliminated by switching the vernier engines to hydrogen, increasing their thrust, and shutting down the main engines at 80,000 feet. Even the existing vernier engines, which are based on the old RL10 design, could be modified to burn high-mixture ratio LOX/ hydrogen. This change would require tankage for the hydrogen that would grow beyond the moldline of the original vehicle, but otherwise would have little effect on the configuration of the first stage.

The other environmental problem aggravated by the massive growth in space operations has been orbital debris. There have been a number of serious impact incidents in the past 10 years, several of which resulted in loss of pressure in either the cargo bay or one of the propellant tanks. Fortunately, no loss of life has yet resulted. Nevertheless, the problem is serious. Already, over half the mass of the low-Earth-orbit space facilities goes into their debris shielding. It had quickly become standard policy to dock all spacecraft within a shielded hangar so that visiting orbital stages would be exposed to the debris environment for as short a time as possible. Early on, Forsyth was active in the drafting and passage of an international treaty laying out strict regulations to control the sources of debris in low Earth orbit.

All low-Earth-orbit satellites are now required to have deorbit systems. Below 400 miles, where atmospheric drag is significant, that can be as simple as a large, inflatable disk that increases the aerodynamic drag sufficiently to bring down a spacecraft in less than a year. The use of aluminum or other metals in solid-rocket-propellant grains was banned early on because the small particles of aluminum oxide can coalesce to form dangerously large particles in orbit. Spacecraft outer surfaces are required to be nonflaking, preferably metal. If not, they have to be stripped on a regular basis, before repeated exposure to the space environment's ultraviolet radiation and free oxygen causes coatings or blanket materials to degrade and break down into loose particles. All tools and small items of equipment for use on orbit must incorporate transponders that can be used to detect them if they happen to float out of open cargo bays or hatches, so that they can be retrieved immediately. Also, because cabin debris tends to collect on or in air filters, any cabin depressurization will cause these particles to come off the filters and float out into space. It has therefore been stipulated that any such filter that could be exposed to vacuum must have a cover that can be closed around it before depressurization.

Regulations were also promulgated to discourage a plethora of small spacecraft in low orbits, which could precipitate a chain reaction of collisions and resulting debris. The small satellite industry was at first strenously opposed to such measures, but as they discovered other opportunities beyond low Earth orbit this became less of an issue. For each orbital altitude, the number of spacecraft is regulated, and as a result the number of larger space platforms and space stations has grown to accommodate the payloads that would otherwise have been on independent free-flyers. The U.S. Coast Guard is in the process of procuring several vehicles for inspection of on-orbit platforms for hazards and to provide the capability to remove defunct spacecraft from orbit.

The major hazard of exploding empty propellant tanks has been greatly reduced by requirements for proper venting of all such tanks and the deployment of electromagnetic braking tethers to bring them down as soon as possible. But more importantly, 10 years after the introduction of the DH-1, there are very few expendable launches. Those that are still being launched are mainly for large, expensive spacecraft that were already far along in the development or production process and not compatible with the DH-1. It is also now standard industry practice to perform spacecraft deployment and other operations that might result in the release of debris at low enough orbital altitudes that the debris will be rapidly brought down by atmospheric drag.

Yes, in 10 years things have really been happening in space. Not only in Earth orbit, but also on the moon, on Mars, and on the asteroids. The level of activity is vastly more than most pundits and prognosticators would have dared to predict back when the DH-1 first came on the scene.

But for John Forsyth, it is still only a beginning.

EPILOGUE II
Mars for the Many

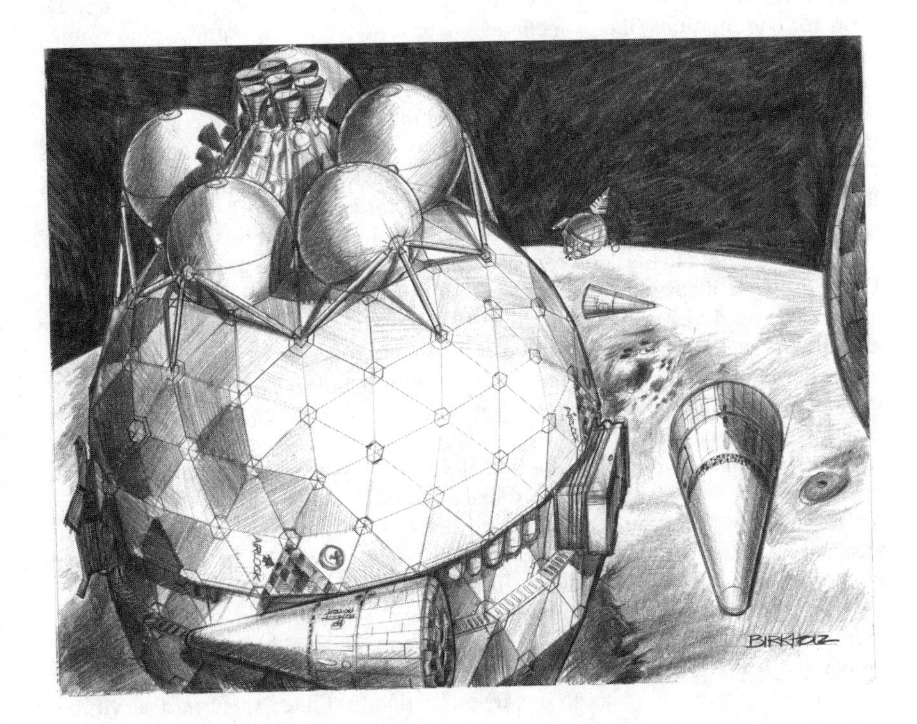

IT was several months later when I returned to AM&M. I was writing a book on the Benedictine monks and their lunar monastery, and I had spent the last few months living at their novitiate in Colorado. I had been able to talk Number 3 into providing me with a ticket to the moon to complete my research there. Before I left on the next monthly supply ship, for what I hoped would be a six-month stay, I wanted to talk to John Forsyth. When I walked into his office, I was surprised to see that it appeared to be in the process of being cleaned out. All of the mementos and awards he had acquired as a result of his accomplishments at AM&M over the last decade were gone; only a few papers remained on his desk.

I knew that Forsyth had been less active in directing day-to-day operations at AM&M these last few years, but it came as a shock when he told me that he was giving up almost all of his responsibilities at the company and would henceforth be working full time to put together a new enterprise, with a Forsyth-typical nondescript name — the Plymouth Foundation. As we talked, it took just a few minutes to see clearly where he was headed. Not content to have merely developed an economic engine that had dramatically succeeded in lowering the cost of space transportation, he wanted to open a real frontier—a frontier with people.

Forsyth had certainly accomplished his goals at AM&M, opening the way for a multitude of human endeavors in space. It's true that not many people are making money in the space transportation business yet, but there have been plenty of industries where that was the case for decades after the creation of the industry. But the economic spiral of lower cost and greater utilization, followed by even lower costs and even greater utilization, is definitely well along. Shipping costs for commodities like propellants are around $100 per pound, and a round-trip ticket to orbit can be had for less than $50,000. There is no regular passenger service to the moon yet, but there are regular cargo flights, supplying the monastery and polar bases, with room for limited numbers of passengers—mostly monastery and the base personnel, but also a growing number of well-heeled tourists. Scheduled passenger flights are expected to begin in a year or two, and a lunar resort for paying passengers is already under construction. It is rumored that the price for a visit, including round-trip spacefare, will be under $1 million.

Forsyth's new focus is on Mars—a planet with a surface area equal to that of Earth, a plentiful supply of carbon dioxide and nitrogen in the atmosphere, and water that could be pumped from the ground in many places. As he explained to me, the pioneering work performed by Father Scipio on the moon has made it clear that, whether or not Mars could or would eventually be terraformed, the technology is available now to support as many people on that planet as you could conceivably get there.

Forsyth made it clear that he is determined to keep Mars from becoming a scientific or environmental preserve like Antarctica. As he sees it, there is a very real danger that Mars could be limited to a few "prestige bases" established by the wealthier nations of Earth, housing a few dozen, maybe a few hundred, men and women who would remain visitors, having only a passing association with Mars. He intends to transport large numbers of colonists to the planet, who will make it their permanent home. In typical Forsyth fashion, he even has a plan for making money while doing it.

As he explained his plan to me, it sounded at first like indentured servitude, with hardships even greater than those endured by the earliest American pioneers. His description of the colony ship brought the term "slave ship" to mind. But as he sketched it all out, it made more sense.

AM&M had never enjoyed the phenomenal run up of its stock value that companies like Microsoft and some of the major Internet companies had seen. Nevertheless, Forsyth had managed to take out over $20 billion, while still retaining a substantial stake. He had done this using the Bill Gates technique of giving the stock to nonprofit corporations, which could then sell the stock, creating a pool of cash while avoiding taxation. Sale of a major stake in AM&M by a nonprofit company did not trigger any concern that Forsyth himself had doubts about the viability of AM&M and its future prospects. Two other investors from AM&M, Number 2 and Number 6, had also committed to the Plymouth Foundation, bringing the total available funds to approximately $30 billion.

THE current estimated cost of transporting a person to Mars and sustaining him there for a year is about $5 million, and so $30 billion was—in theory—enough to transport 6000 people to Mars and support them there for a year. However, using all of the foundation's capital for transportation wouldn't leave anything for growing the colony. Forsyth therefore realized that he needed to make it a paying proposition, or a nearly paying proposition, so that many thousands of colonists could be transported to Mars, and the infrastructure needed to open up a whole new world and build a new civilization would be available. Even though the Plymouth Foundation was a nonprofit corporation, Forsyth intended to generate as much income as possible to gain more leverage from his money.

He had undertaken, with help from the advanced planning staff at AM&M, to investigate the absolute minimum cost of transporting a person to Mars. Making a basic assumption that the oxygen-carbon dioxide cycle and the water cycle could be closed with about 98% efficiency by employing relatively simply electrochemical systems, one needs to consider primarily the mass of the person and the food that they will require. That includes food for the trip out, plus the weight of 1000 square feet-worth of greenhouse for use on Mars, and six months' worth of food that they can eat while the greenhouse is built and until the first crop can be harvested. Of course, as the colony grows, the materials for the greenhouse and the initial food supply will come from local sources, eliminating the need to ship them out from Earth and greatly reducing transportation costs.

The AM&M engineers, in consultation with nutrition experts, have concluded that 1 pound of high-fat-content food per day, providing somewhere between 1600 and1800 calories (depending on the size of the individual colonist), will be sufficient for what will be a relatively low-physical-activity trip. An initial load of 3 gallons of water per person will be needed to start the recycling process. The total weight budget per person will be 150 pounds for their own (average) body weight, 25 pounds of personal possessions (including clothing), 25 pounds of water, 270 pounds of food, and another 1000 pounds for the six-month food supply, greenhouse materials, and other equipment. That comes to about 1500 pounds per person for the initial waves of colonists, dropping to around 500 pounds after the point when newcomers can be fed from Martian sources. With propellants still coming up from Earth, and about 3 pounds needed for every pound transported to Mars, the total mass needed in low Earth orbit will be some 6000 pounds per person initially and about 2000 pounds when food and other supplies are available on Mars.

To achieve minimum costs, a cycling spaceship of the type that enters into highly elliptical orbits at each end of its cycle, massing nearly 1000 pounds per person transported, will be needed to reduce the recurring weight budget to near the theoretical limit. The spacecraft that the AM&M advanced projects team has designed for Forsyth consists of a 100-foot diameter sphere, providing just about the most cramped living quarters imaginable. Each passenger will be allowed to leave their compartment for 1 hour of exercise and 1 hour for eating and personal hygiene per day. The ship will be rotated so that most of the accommodations have Mars-normal gravity. The engineers figured that a dual-redundant nuclear power system with a total capacity of 2.5 megawatts will be sufficient to supply the power needed for air and water regeneration, lights, air conditioning, and Internet access for all.

At a cost of $150 a pound to low Earth orbit, the one-way ticket price would in theory need to be $900,000 initially, and $300,000 after the colony is established. There are plenty of people on Earth who have a productive output of over half a million dollars a year and who carry mortgages of over a million dollars, particularly on the east and west coasts of the United States. The Plymouth Foundation team has calculated that in the first decade, 100 pounds of equipment per capita per year will need to be imported from Earth. At $600 per pound delivered to the Martian surface, that will cost $60,000. The mortgage on each colonist's transportation cost of $300,000 will come to about $15,000 annually. Thus each settler will need to produce only about $75,000 in income over and above the effort necessary to develop and sustain the Martian infrastructure and economy. Assuming a 50% tax to support these costs, that will require a per capita productivity of about $150,000.

The cost of supporting the entire first ship's population would be about $500 million a year, which Forsyth believes can be earned—at least in part—by working a variety of jobs at the various national Martian research bases. It has always been difficult to pin an exact value on an hour of an astronaut's time in orbit, but Forsyth believes that skilled, reliable workers on Mars, able to design and build facilities for the bases and provide support to a variety of scientific programs, would be cheap at $50,000 per day. And, undoubtedly, there will sooner or later be private expeditions on the planet that will be willing to pay technically trained "sherpas" reasonable daily fees.

Converted terrestrial transport vehicles, a nuclear power plant, materials for the greenhouses, a six-month food supply, chemical air regeneration equipment, a small chemical factory that will be able to produce everything from liquid hydrogen, liquid methane, and liquid oxygen to a form of dynamite, and a large assortment of hand tools—shovels, picks, drill rods, and everything necessary for a well-equipped blacksmith's shop—will all be prepositioned on the Martian surface. There will be a small colonial administration staff of about 100 people, who will provide the training and direction necessary to get the greenhouses up quickly and provide living space for the colonists who go out in the first ship.

The cycling spaceship will have a delta-V capability of up to 3000 feet per second, but only 40% of that is needed to escape from a highly elliptical Earth orbit. It actually takes more velocity to return from Mars to Earth because of the smaller gravity well around Mars and the consequently smaller boost you get from swinging around that planet. On the other hand, the spacecraft will be about 1/3 lighter, because it will be coming back essentially empty. Using minimum energy transfers, the round-trip travel time from Earth to Mars and back will take about three years, plus another year of waiting until Earth and Mars are aligned for the return trip. Using a single spacecraft, 5000 people could be transported to Mars every four years—an average yearly migration of 1250 people. Assuming a 2% population growth rate—actually quite low, considering that almost all of the colonists will be of childbearing age—over a million people could be on Mars by the end of the first century of colonization, using just a single ship. Forsyth's intent is to be sure that with the money he has available, a sustainable colonization effort of at least that magnitude will be set in motion.

... over a million people could be on Mars by the end of the first century of colonization, using just a single ship.

To my mind, this all seems a rather fantastic scheme, but Forsyth told me that Father Scipio is already planning to join him in the enterprise by founding a monastery on Mars. Considering what John Forsyth has already managed to accomplish, and with the continued growth of the world's economy, it is hard to dismiss his prospects for achieving this goal as well.

Standard Characteristic for Analysis of the Production DH-1

DH-1 Parameters	1st Stage Boost w/o 2nd Stage	2nd Stage 6 RL10A-4	2nd Stage 2 RL-60	Planetary Stage 1 RL-60
GLOW	209,000 lb	99,000 lb	99,000 lb	103k lb- 217k lb
Isp S.L.	260 s			
ISP Vac	310 s	444 s	444 s	465 s
Burn out weight	40,000 lb	17,000 lb	17,000 lb	21,000 lb
with payload	w/o payload	18,000 lb	18,000 lb	135,000 lb
Propellants M/R	Lox:CH_4 4:1	Lox:H_2 6:1	Lox:H_2 6:1	Lox:H_2 6:1
Fuel Wt.	34,000 lb	11,700 lb	11,700 lb	11,700 lb
Fuel tank Vol.	1,300 ft^3	2,775 ft^3	2,775 ft^3	2,775 ft^3
Oxidizer Wt.	135,000 lb	70,300 lb	70,300 lb	70,300 lb
Oxidizer Vol.	2,000 ft^3	1,400 ft^3	1,400 ft^3	1,400 ft^3
Thrust/Weight	1.6	1.35	1.3	0.6–0.3
Payload Vol./wt.	99,000 lb	1,100 ft^3 5,000 lb	1,100 ft^3 5,000 lb	4,000 ft^3 40,000 lb
Total engine Wt.	6,000 lb	2,220 lb	2,200 lb	1,100 lb
Manned Systems & GNC Wt.	2,000 lb	1,000 lb	1,000 lb	5k–20k lb
Structure/Tanks	10,000 lb	6,600 lb	6,600 lb	8,000 lb
RCS	1,000 lb	500 lb	500 lb	500 lb
Residuals	1,000 lb	400 lb	400 lb	400 lb
Recovery Systems Wt.	20,000 lb	1,500 lb	1,500 lb	Rocket Landing
DH-1 Dry Wt + Res.	20,000 lb	12,220 lb	12,200 lb	12,220 lb
Reentry Base Heating	150 BTU/ft^2	2,500 BTU/ft^2	2,500 BTU/ft^2	1k–5k BTU/ft^2
Radius/Heat Shield Wt.		32 ft/700 lb	32 ft/700 lb	32 ft/1000 lb
Stage Height	30 ft	44 ft	44 ft	64 ft
Cone Angle	none	11.5°	11.5°	11.5°/14 ft Cyl.
Base Diameter	25 ft	20 ft	20 ft	20 ft

Index

Supporting Materials

A complete listing of AIAA publications is available at http://www.aiaa.org.